Pvt. Ltd.

A useful Guide for and Research Scholars perusing PhD course work and for Faculty and Professionals in the field of Science and Technology

Communication Skills in Scientific Research

Munish Garg
&
Chanchal Garg

Department of Pharmaceutical Sciences
Maharshi Dayanand University
Rohtak, Haryana – 124001

2017

Studium Press (India) Pvt. Ltd.

Communication Skills in Scientific Research

ISBN: 978-93-80012-99-5

Published by:

Studium Press (India) Pvt. Ltd.
4735/22, 2nd Floor, Prakash Deep Building
(Near Delhi Medical Association),
Ansari Road, Darya Ganj, New Delhi-110 002
Tel.: + 91-11-43240200-15 (15 lines); Fax: 91-11-43240215
E-mail: studiumpress@gmail.com

Printed in India

About the Authors

Dr. Munish Garg, Doctorate in Pharmaceutical Sciences is presently working as Professor and Head, Department of Pharmaceutical Sciences, Maharshi Dayanand University, Rohtak, Haryana. He has about 15 years of professional experience in teaching and research. He has published more than 50 scientific papers and 100 scientific abstracts in the journals of repute. He has published 4 patents and 2 books to his credit. He has also presented his work at different platforms of National and International level. He has received sizeable amount of funding from different Government organizations.

Dr. Chanchal Garg, Doctorate in Pharmaceutical Sciences is presently working as Post Doctorate Fellow (awarded by University Grants Commission, New Delhi) in Department of Pharmaceutical Sciences, Maharshi Dayanand University, Rohtak, Haryana. She has about 10 years of professional experience in teaching and research. She has published about 10 scientific papers in the journals of repute. She has also 2 book chapters and one patent to her credit.

Acknowledgements

We would like to extend my sincere gratitude to the University authorities especially Honorable Vice Chancellor Prof. B. K. Punia and Registrar, Jitender Bhardwaj for being the source of inspiration, which made it possible for us to write this book.

We would like to extend our thanks to all my colleagues who have been helpful in bringing this book to its final shape.

Our heartiest thanks are due to all the research scholars especially Saurabh Satija, Meenu Mehta, Meenu Bhan, Hitender Sharma and Pritam Singh who have extended all possible help on a single call.

Thanks to our sons Aditya and Aarush for their unconditional love and support.

Finally, thanks to the Publishers for their continuous support and valuable suggestions throughout the project.

Munish Garg
Chanchal Garg

Preface

Scientific Communication skills are essential for every researcher. They must be able to explain their work effectively and memorably to non-specialist audiences as well as to their peers. It is the heart of science, both in making claims to new knowledge and transmitting the body of knowledge thus, is an integral part of scientific research. The persons, who are entering a scientific profession, must be familiar with fundamental communication skills and knowledge to communicate and excel in the sciences. This book is written with an objective to help the professionals who are new to the profession as well as those are well experienced and need to sharpen their communication skills. The chapters in this book are well designed as per the needs of the curriculum of PhD course work as well as working professionals.

First chapter is about introduction and types of scientific communication. Next three Chapters explain importance of publishing a paper, components of a scientific paper and publication process. How to write a review article is discussed in detail. Scientific oral and poster presentation is an important activity in a researcher's life so these two aspects are explained in detail in chapter five and six. Plagiarism issue which is important aspect in scientific writing has been discussed in sufficient depth. To get the financial grants from different funding agencies is a dream as well as need of every researcher. This chapter explains different funding agencies, their schemes and guidelines to prepare a proposal are given in detail.

Few sample proposals are also given for further help of the researchers. Next chapter is related to intellectual property rights and patenting aspects which are necessary these days and proper awareness among researchers is a must so each aspect is written and presented in detail. The technology transfers, MoU, confidential agreements which are necessary at industrial scale are elaborated in this book. Further every researcher need basic guidelines to use common computer programs for research and communication which are written in very detail. After reading this chapter even a novice can work properly in these programs. Concept of internet and basic search engines are also written and presented in brief in this book.

We would like to extend my sincere gratitude to the University authorities especially Honorable Vice Chancellor Prof. B. K. Punia and Registrar, Jitender Bhardwaj for being the source of inspiration, which made it possible for me to write this book.

We would like to extend our thanks to the Dean of Faculty Prof. B. Narasimhan and all my colleagues who have been helpful in bringing this book to its final shape.

Our heartiest thanks are due to all the research scholars especially Saurabh Satija, Meenu Mehta, Meenu Bhan, Hitender Sharma and Pritam Singh who have extended all possible help on a single call.

Thanks to our sons Aditya and Aarush for their unconditional love and support.

Finally, thanks to the Publishers for their continuous support and valuable suggestions throughout the project.

Munish Garg
Chanchal Garg

Table of Contents

1

Basics of Communication Skills

> *"Nothing in science has any value to society if it is not communicated, and scientists are beginning to learn their social obligations"*
>
> ***Anne Roe, The Making of a Scientist (1953)***
>
> *"We need to apply the science of communication to the communication of Science"*
>
> ***Preston Manning***

INTRODUCTION

Scientific communication is sharing the ideas, findings and reviews of a scientific professional or scientist to the world in the form of writing papers in the journals, oral and poster presentations.

It is the heart of science, both in making claims to new knowledge and transmitting the body of knowledge to both Scientists and non-Scientists thus, is an integral part of scientific research.

Have you ever recognized how many brilliant ideas, thoughts and theories get unnoticed just because no-one ever heard about them, probably millions. Did you ever wondered, why Dr XYZ's, common findings are cited in many good journals and your amazing result goes unpublished. The answer is because of the communication that is lacking. Good communication is an essential tool if you want your ideas to flourish and fly.

The study and practice of science has its own procedures and protocols as well as an oral, written and visual language. A Science professional who is entering a scientific profession must be familiar with fundamental skills and knowledge so that he can communicate and excel in the sciences.

As a scientific communicator it is expected that he should be able to:

- Demonstrate the ability to write a clear and concise scientific report.
- Demonstrate the ability to do oral or poster presentations to a variety of audiences.
- Demonstrate the ability to work in a group.
- Demonstrate an understanding of basic computer software used in the sciences.
- Be able to effectively search the web for relevant data.
- To communicate for getting funds from a Government/private source.

TYPES OF COMMUNICATION SKILLS

Pictorial representation of types of communication skills is given in Fig. 1.1.

Professional Science Communication

This type of communication is the most formal type of scientific

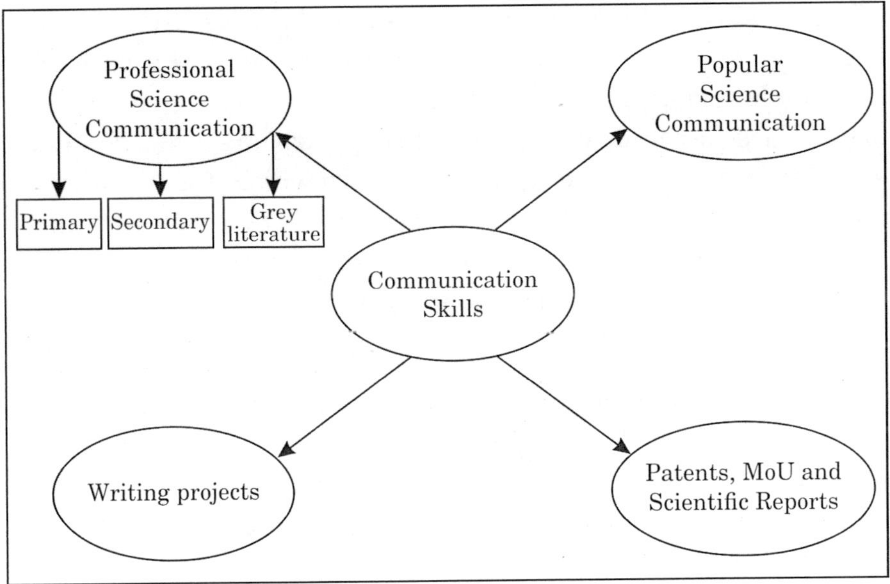

Fig. 1.1: Pictorial representation of types of communication skills

communication made from scientist to scientist. It normally leads to a formal publication of the results, findings, observations and views arising from a scientist's research project. Most often, the official results are published in the form of printed materials, such as academic journals. Verbal communication channels, such as personal contacts with colleagues and teachers, seminars, workshops, lectures, and conferences, are also vital to the exchange of information among scientists. These types of communications work toward the advancement of the various scientific disciplines.

Different types of scientific literature exist, normally referred as,

- Primary
- Secondary
- Grey literature

The ***primary literature*** refers to accounts of research carried out personally by an individual scientist or as collaboration by a group of scientists, which is published in a peer-reviewed scientific journal. These accounts commonly called 'papers' are written in the particular format specified by the journal to which it is submitted for publication. Most primary literature is published in scholarly journals, but some research is published as monographs, theses or dissertations, conference papers and reports.

The ***secondary literature*** consists of publications that rely on primary sources for information. Here it is not a requirement for the authors to have done the work themselves, since the purpose of the publication is to summarise and synthesize knowledge in a specific area for other scientists who already have an understanding of the topic; however, the authors of secondary publications would normally have worked and published primary literature in the area they are writing about. The secondary literature includes review journals, monographic books and textbooks, handbooks and manuals. Although normally written in a scientific style, secondary publications are not organised in the same way as the primary publications are; however, it is a universal requirement that they are fully referenced and that most of these references are to the primary literature.

Grey literature refers to sources of scientific information that are not published and distributed in the usual manner and which therefore may be difficult to obtain. Grey literature includes theses and dissertations, technical reports with a limited distribution, journals published by special interest groups that have a limited distribution, abstracts of conference papers and conference proceedings that are only made available to conference participants. Grey literature here does not mean that this literature is in any means has less scientific merit, but the distribution and access of this type of content is limited.

Popular Science Communication

This type of communication is made from the Scientist to non-specialized audience. It generally refers to public media discussion about science topics to a non-scientist, general audience. This audience can be composed of children, teenagers and adults. Often, scientists are involved, in order to ensure the correctness of the information transmitted; but the communication is done in terms that the general public can understand. Scientific communication can be done through events, television programs, journal and magazine articles, as well as science-related programs and policies.

Writing Projects

This type of communication involves the skills of a scientific professional to communicate with various Government and private funding agencies to get sufficient financial grant. Every agency has its own formats and priorities so a good communicator should be able to represent his project in an effective and appropriate manner to get the grant. Sometimes it becomes necessary to communicate to get grants for travel and various scholarships.

Patents, MoU and Scientific Reports

The communications are made to get a patent or memorandum of understanding or for technology transfer. The documents for these types of scientific communication are prepared very carefully due to legal conditions. The scientific reports are also prepared in this type.

2

Importance of Publishing a Paper

FORMS OF PUBLISHING A PAPER

Publishing a paper is an integral part in the life of a researcher. Not only it gives recognition to the researchers but sometimes the salary, promotions etc. are also linked to the researcher's professional achievements those are mostly in academics. Writing a paper may not be the hobby or liking of a professional because of a tedious and time consuming job but still every academic professional wants good number of publications to his credit. Writing in a scientific form may be difficult for a new comer but this can be learnt with time and experience. The scientific paper can be published in the following basic forms:

- Research paper
- Review paper
- Publication in conference proceedings
- Book Chapters

Research Paper

Research papers are one of the scientific communication forms in which new findings obtained through a set of experimental studies by a researcher are communicated for publication. A suitable journal which accepts the kind of work carried by a researcher is selected and according to their required format, the paper is written and communicated. The paper is generally reviewed by reviewers and also checked by language experts. If suitable, the paper is accepted for publication in that particular journal.

Review Paper

Review paper is not based on original new findings but a researcher who has thorough understanding of the subject reviews a set of papers published already on a particular topic. The researcher critically examines the findings, correlate and presents the view so that other researchers can easily understand the topic and plan their future research accordingly. Although no experimental work is done to write a review paper but still requires a lot of time to collect data, compile, examine and present in a simple and effective way. Review papers generally bear the same weightage as that of research papers.

Publications in the Conference Proceedings

In seminars, conferences, symposiums and congress programs the researchers participate and present their finding through oral or poster presentations. Before presenting the work in such programs the abstract of the work is communicated to the organisers and after due review process the suitable papers are accepted and invited for presentation. The presented papers are then published in the form of conference proceedings. In the proceedings the full length papers are prepared by the researchers and communicated. It is the same manner as the paper is communicated for publication. Although the conference proceedings are meant for limited distribution, this type of work remains as grey literature (as explained in previous chapter) but in no way less important than the original published work. Again it depends upon the type of conference, the review system adopted to select the papers etc. Proceedings of some of the top-ranked conferences are equally or even more prestigious than articles published in highly ranked journals.

Book Chapters

Book chapters are also contributed by individual professionals or in a group. The chapters are invited by the editor or editors of a book from the professionals who are expert in their respective fields. The book chapters are contributed by different professionals and the whole book is prepared by the editors and published by the publisher.

IMPORTANCE OF PUBLISHING PAPERS

The publication of papers is very important for a researcher. Every researcher tries to gather good number of publications in his life. There are various but obvious reasons behind this as detailed below:

Approved Work

Journal articles are the approved work of a professional. When a journal accepts an article for publication, it means the work carried by the researcher is worth publishing. This approval is communicated to the wider community that the work carried out has a particular importance as judged by the journal.

There are different types of journals in which one wishes to publish his work. Few Journals have double blind review process, while others may have triple or single blind review process. It is also not necessary that if a paper is rejected in any given journal is not worth publishing in other journal. Every journal has its own priorities so it is quite possible that a particular paper does not fall in the journals domain.

It is not necessary that criteria for selection of a paper for publication will be the same for every journal. The review process followed by journals is different in each case and thus the results also may be different. Journal x may be double-blind while journal y is triple-blind. Or journal x may use one referee while journal y will use two or three referees. There are also differences in persons used. Some journals may have a more prestigious list of editorial board members and referees to use in assessing essays than another. Therefore, even if two journals had the same formal procedures for assessing essay submissions, papers would be assessed by academics with very different backgrounds.

Record of Research

Publishing a paper is a record of work carried out in a particular field. It helps the community to plan their research and cumulative research publications result in a significant discovery.

Legal Rights

A published work of a researcher is to register the discovery on a certain date. Without publications, it would become difficult to give credit to the responsible professionals. It also protects the legal and intellectual property rights of a researcher. Publication leaves a permanent record of a research.

Acquiring Degrees

In most of the universities, publishing of papers out of the PhD research project is a requirement for the award of PhD degree. Thus it becomes necessary for the candidate to publish the stipulated number of papers to complete the degree. Sometimes, publishing of paper/s is also mandatory for the award of even postgraduate degree.

Economic Aspects

Publishing chapters in books are written and contributed by different authors by invitation as mentioned above. Sometimes all the chapters in a book are written by same authors. As the published books are sold to the researchers and professionals on payment basis, a royalty amount is given to the authors/editors of the book. So, publishing of papers results in economic advantage to the authors. The amount paid as royalty to the authors/editors depends upon terms and conditions finalized between authors and publisher at the time of publishing of a book. Few journals also pay some amount to the authors for publishing papers in their journal.

Recognition

Publishing paper brings recognition to the authors and results in several invitations to the author/professional to deliver lecture at national and international platforms. The professionals having good number of published papers are invited to review papers of reputed journals and their names are included in the list of reviewers or in the editorial board. Various funding and grants to be given to a professional are based on the number and quality of published work. Also based on their good publications the professionals are included in various review committees, expert committees and Government panels for different activities. Thus publishing papers bring sufficient recognition in the field.

Feedback and Scope for Better Future Work

Researcher gets feedback from readers which prepare him to do better work. The published work of a professional is read by the whole scientific community so critical views are received by a researcher to enhance the skills.

Academic Job Qualification

In large part due to the fact that journal articles are often considered as base for academic jobs as they are seen as academic job qualifications.

The journal articles can also play an additional role in providing further job qualifications both to acquire a new position and to earn promotions.

3

Components of a Scientific Paper

CHOOSING A JOURNAL FOR PUBLICATION

Before writing a scientific paper, it is very necessary to identify or chose the journal in which the paper is desired to be communicated. The journal for communicating the paper for publication must be selected carefully as right selection of journal also increase the chances of paper acceptance. Every journal has its own priority area and a particular kind of papers is published. For example, Journal of Ethnopharmacology accepts the papers in which traditional drugs are scientifically screened for a activity. The papers with more chemistry work, isolation work etc. do not get place in this journal. But papers having more chemistry work are preferred in Journal like, Planta Medica. So it becomes necessary to select the journal according to the type of study conducted. After the selection of particular journal, the paper is written accordingly because every journal has its own style of writing.

These days many new journals have emerged which are having ISSN numbers but not have any established indexing like Pubmed, Scopus etc. Many journals require processing fee also for publication of articles. So before submitting the paper for publication, these things must be considered.

REFEREED JOURNALS

Refereed journals are also known as peer reviewed journals. It means the matter before publication in the journal is reviewed by the editors as well as reviewers to check the quality and give their recommendations whether the article is suitable for publication or not, or whether it needs some improvements before publication. This exercise is done by most of the journals to ensure the quality of the article. The review process may be single blind, double blind or triple blind as discussed in later sections of this chapter.

- To check whether a journal is peer reviewed or refereed or not, one must go to the ulrichsweb.com (AKA Urlich's International periodical directory) and search for the desired journal. If the name exists in this directory, the journal is refereed.
- Secondly, in the author's instructions of the journal, the journal always discloses whether there is any review process or not.

INDEXED JOURNALS

Often the indexing of a journal is checked by the readers and article contributors. The journal also tries to sell itself by displaying indexing of the journal by various agencies. Whenever the website of a journal is visited, the details of indexing are found. But one should be careful that all indexed journals need not of good quality. Scientific citation index (SCI) maintained by Thomson Reuters is considered quality indexing and journals finding place in this index are considered to be of good quality.

IMPACT FACTOR

Impact factor of a journal reflects the average number of citations of articles published in the journal. Any paper published in a journal is referred by others and citations are made in their papers. These citations are counted every year for every paper and every journal and overall average number is calculated. Higher the number, greater the reputation of the journal is considered. This number is often used to judge the quality of a journal. Although there are many controversies over this issue whether considering only impact factor is a justified criterion to judge the quality of a journal or a paper but still mostly this fact is accepted.

For calculation, for any given year, the number of citations is calculated for a paper in the next two preceding years. For example the impact factor of a journal in 2013 is 4 then it means that 2011 and 2012 the average number of citations of its papers was 4. New journals which have started their publications but not indexed will start getting their impact factor two years after the indexing. The journals which are indexed from day one will get the impact factor after two years of indexing.

It should be clearly understood that most of the journals do not have impact factors and they are not even having acceptable indexing. So, one should be very careful about misleading terms like proposed impact factor, a long list of indexing which is not acceptable and ICV values which are often displayed in many journal websites to sell their journals. These are not at all impact factors. Impact factor of a journal can be found on SCI website maintained by Thomson Reuters by the name Intellectual property and Science.

h-INDEX

The h-index also known as Hirsch Index suggested in 2005 by John. E. Hirsch is an index that attempts to measure the productivity or citation impact of the published work of a researcher, a group of scientists, an organisation or a University. The index is based on the set of the scientist's most cited papers and the number of citations that they have received in other publications. For example a scientist having h-index 5 is having 5 published papers which are cited at least 5 times. So this index tells about both the number and number of citations. h-index is considered these days as an indicator to judge the quality of an individual or an organisation.

STRUCTURE OF A PAPER

The general structure of paper has several components. A schematic diagram (Fig. 3.1) is presented below to show the simplified structure of a paper.

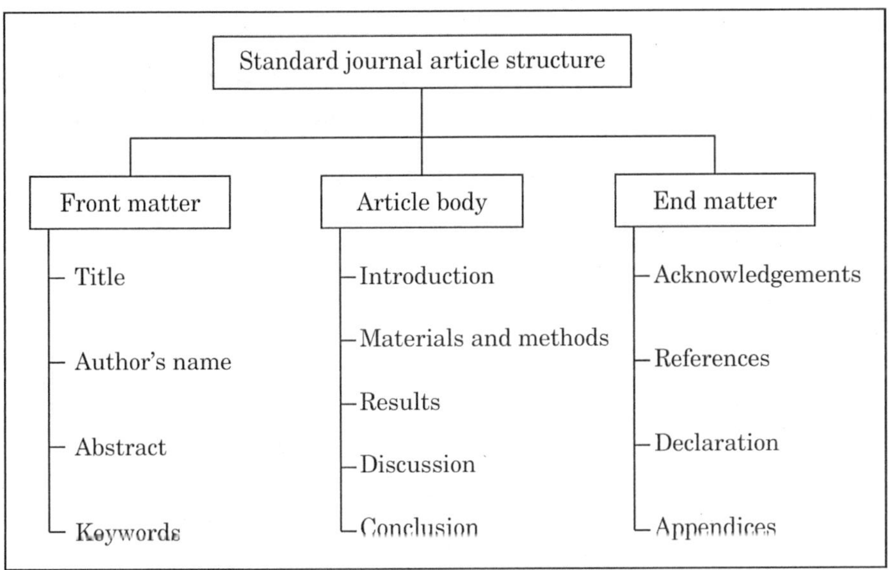

Fig. 3.1: Components of a standard article structure

Title

Writing title of the paper is the first and most important part. The title is written in such a way that it should be able to stimulate the reader's interest. The purpose of a title is to give the reader a guess of what the paper is about. Both very short and very long titles are not recommended. The title should be of optimum length. It is the gateway to a paper. Many places the paper is judged by title and if the title is not impressive then the paper is not further read and straightaway rejected. The characteristics of a good title are:

- Accurate
- Informative
- Descriptive
- Simple
- Specific

The title of a paper should be written in such a way that it should avoid the filler words like, studies of........ Title of the paper should not contain any abbreviations and should not be in the form of a question. There should not be any spelling mistakes and title should be such that it gives sufficient information of the work explained in the paper. It is thus advisable to write 3-4 titles and then decide the best one.

Author's Details and Affiliation

The details of the authors and their affiliations are written just below the title of the paper. The author's names should be written as per journals requirements. Few journals require full name like, Suresh Gandhi while some journals require short name like, Gandhi S. The name of the authors who have sufficiently contributed in the work should be included and mere post or designation like head of the department does not entitle anybody to include his/her name in the authors list. Thus the sequence of names of the authors to an article must reflect the relative scientific or professional contribution of the authors, irrespective of their academic status.

There is no problem in case of single author but when there are two or more authors and the number is further increased then it becomes a question to decide the sequence in which the name should be written.

As a general rule, the name of the principal contributor should come first with subsequent names in order of decreasing contribution but the name of the guide, supervisor, and mentor should reflect at the last.

A student should be listed as a principal author in a multi-authored article that is substantially based on the student's dissertation or thesis.

The proper address in the affiliation of authors should be written individually and should be written with full details or as per the requirement of the journal.

Writing Abstract

Abstract is a short summary of a paper written before introduction. It can be structured in which the various parts like objective, materials and methods, results, discussion and conclusion section are formed or can be non-structured in a paragraph but the matter is written which covers all above mentioned headings. This is the major portion which helps the editors to make decision about a paper to read further or to consider the suitability of the paper for publication. The readers often go through the abstracts for their literature survey and select the abstracts for further reading. A good abstracts helps to get good citations while a bad abstract do not. The abstract should be clear, honest, and concise and should cover all the major points. Various types of abstracts are as follows:

1. ***Descriptive abstract*:** or topical abstract describes the contents but contains too little substance and detail.

2. ***Informative abstract:*** Self-explanatory report on a scientific investigation – 200 to 250 words.
3. ***Extended abstract:*** Conference proceedings

Selecting Keywords

Keywords are main 4-5 words written after the abstract. These words help to search paper. For example if someone searches for antihypertensive plants then the papers with matching key words will be shortlisted and will appear as search results. A researcher searches the topic like this and the papers which are referred are having chances to get maximum citations. So key words must be selected carefully that the words which are most relevant to the study and the written paper must be selected. For example somebody has carried out HPTLC analysis of a plant *Sesbania grandiflora* then surely two main key words *Sesbania grandiflora* and HPTLC should be included in the list of keywords.

Introduction

Introduction is written after the abstract and this portion leads the reader from broad subject area to a specific aim. This is the part where an author justifies why the study was undertaken and on what basis the study was designed and planned. In a research or review paper the introduction section is of 2-3 paragraphs and contains sufficient references. There are three main elements of introduction,

Background of the subject

In this section the background of the subject area is defined and work done so far in the field is explained with proper references. Mean to say that in section an author try to explain what is known till date.

Establish a gap / Problem

In this section the problems and gaps are identified. The gaps in the previous studies are identified and establish the need to fill those gaps.

Occupy the gap / Problem

In this section a proposal of the study planned is made and the author explains the effort to occupy those gaps identified in the previous section. The justification for why the present study was conducted is given.

Materials and Methods

This is very important section of a paper because it provides the information that how the study was actually conducted and the validity of a study is ultimately judged. Therefore the author must provide a clear and precise description of how an experiment was carried out and the rationale for the specific experimental procedures chosen. It must be written with enough information so that:

1. The experiment could be repeated by others to evaluate whether the results are reproducible or not.
2. The audience can judge whether the results and conclusions are valid.

This section is mainly divided into two parts where the materials section refers to what was examined (*e.g.*, Humans, animals or tissue preparations) and also to the various treatments (*e.g.*, Drugs, gases) and instruments (*e.g.*, Ventilators) that were used in the study. Also what aids were used like chemicals solvents and their source of purchase. "Methods" referred to how subjects or objects were manipulated to answer the experimental question, how measurements and calculations were made and how the data were analyzed. The complexity of scientific inquiry necessitates that the writing of the methods be clear and orderly to avoid confusion and ambiguity. First it is helpful to structure the methods section by:

1. Describing the materials used in the study
2. Explaining how the materials were prepared
3. Describing the research protocol
4. Explaining how measurements were made and what calculations were performed
5. Stating which statistical tests were done to analyze the data.

Second, the writing should be direct and precise and in the past tense. Compound sentence structures should be avoided as well as descriptions of unimportant details should be avoided.

Results

Results section is an explanation of the observations and data generated out of the experiments conducted. It is logical display of data and findings. The results are often described in the form of tables, figures, charts so that one should be able to understand the data easily. Suitable statistical analysis is carried out and displayed. The results are also compared with previously conducted studies. Sometimes result section is also combined with discussion section.

Tables and Figures

As mentioned above, it is good to represent the data in the form of tables and figures. But few journals limit the number of tables and figures so these should be prepared accordingly. The data represented in table or figure should not be repeatedly explained in the text. Figures can be displayed in the form of pictures, flow charts, pie charts, line charts and bar charts. Various softwares are available now-a-days which help to prepare them easily. The title of the table is written on the top of the table while the details of the figure are written below the figure called legends.

Discussion

The purpose of the discussion is to interpret the results and correlate with objective and previously conducted studies. The gaps identified in the introduction sections are filled in this section. The implications of findings are explained and suggestions are made for future research. Its main function is to answer the questions posed in the introduction, explain how the results support the answers and how the answers fit in with existing knowledge on the topic. The discussion is considered as the heart of the paper and usually requires several writing attempts.

1. The discussion is organised from general to very specific.
2. It is written in present tense.
3. Begin by background information and recapitulating the aims.
4. Support the answers with the results.
5. Discuss and evaluate conflicting explanations of the results.
6. Summarize concisely the principal implications of the findings regardless of statistical significance.
7. Provide recommendations (not more than two) for further research.
8. In writing the discussion, discuss everything but be concise, brief and specific.

Conclusion

Conclusion is a short but most specific statement given at the last to present overall impact of the study conducted. Conclusion is stated clearly and precisely. At the end, a short statement regarding the significance of the work is given.

Acknowledgements

Remember to thank the funding agency and colleagues/scientists/ technicians who have provided assistance

Different parts of an acknowledgement are as follows:

- Financial (recognition of extramural or internal funding);
- Instrumental/technical (providing access to tools, technologies, facilities and also furnishing technical expertise such as statistical analysis.
- Conceptual (source of inspiration, idea generation, critical insight, intellectual guidance, assistance of referees etc.
- Editorial (providing advice on manuscript preparation, submission, bibliographic assistance etc.)
- Moral (recognising the support of family and friends etc.)

References

References are the citations made in the text about the previously conducted studies and writing their full bibliographical details at the end of the paper. This is very important part of scientific writing and has lots of significance. The correct selection of references and using them at appropriate place make a paper impressive and scientifically strong. There are different styles of writing references. Every journal requires its own style of writing references. But regardless of the citation style, there are two basic rules for the list of references: (1) every cited source must be listed, and (2) every listed source must be cited. Many different styles of referencing have developed over the years. Currently there are three main styles of citations in the text:

Name and year system

In this system, the references are cited by authors name followed by year. For example, Gupta and Shukla (2003) have...... This is a very convenient system of citations as there is no problem when the citations are added or removed. But the problem arises when there is more number of citations in a single line because it becomes difficult for the readers.

Alphabet-number system

In this system, the references are shown alphabetically and citations are made by the numbers. Relatively it is easy for the reader as it does not break the flow of reading.

Citation order system

In this system, the references are arranged in the manner of their citations and citations are made as numbered in the order of their appearance like, Hypoglycemic drugs[1]. The numbers can be shown in different ways *e.g.*, as superscripts, brackets etc.

There are variable systems of referencing which depends mostly upon the journals choice or the type of scientific discipline. Few are detailed below:

1. ***The APA style:*** This System is also known as the Harvard style or name (date) system because in this system, author's surname in the text is followed by the date of the publication in brackets and entries in the reference list are listed alphabetically, starting with the name and the initials of the author followed by the date of publication for each entry.

   For example: Garg, M (1998). Ionic Liquids in modern extraction of bioactive compounds. In P. Sharma & V. Gupta (*eds.*), Novel Extraction tools of Medicinal Plants (pp. 125-40). New Delhi: Daya Publishing House.

2. ***The Modern Languages Association (MLA) style:*** In this system, the author's surnames appear in the text and the first

author's surname comes first in the reference list. This is followed by names of other authors. Dates of the publications are given after journal titles or at the end of the references for books etc. The list is ordered alphabetically.

For example: Singh P, Verma P. Antidiabetic activity of *Moringa oleifera*. Journal of Ethnopharmacology 4 (2011) 111-14.

3. ***The Institute of Electronic and Electrical Engineers (IEEE) style***: In this sytem, the authors in the text are numbered in order of their appearance in the text. The reference list is then numbered sequentially. Names are presented with the initial first followed by sirnames. Dates of the publications are given after journal titles or at the end of the references for book etc. For example, P.K. Shastri, Antidiabetic activity of *Moringa oleifera*. Journal of Ethnopharmacology vol 3, no.2, pp. 111-14, 2011.

4. ***The Vancouver style***: This system is very popular in scientific writing. Similar to the IEEE system, the authors are numbered in the text in order of their appearance and numbers are enclosed within the brackets. The reference list is numbered sequentially but the authors are listed surname first followed by their initials. Again the dates of publications are given after journal titles or at the ends of the references for books etc.

   For example, Shastri P.K., Antidiabetic activity of *Moringa oleifera*. Journal of Ethnopharmacology. 2011, 3(2), 111-14.

PUBLICATION PROCESS

Every journal has its own process of reviewing before publishing a paper (Fig. 3.2). Few journals have single blind review process while others have double or triple blind review process. Every journal has its own priority area in which the papers are published. Any given paper rejected by certain journal does mean that the paper has poor quality. The same papers may be sent to another journal but before sending the paper anywhere for publication, the priority of the journal must be gone through.

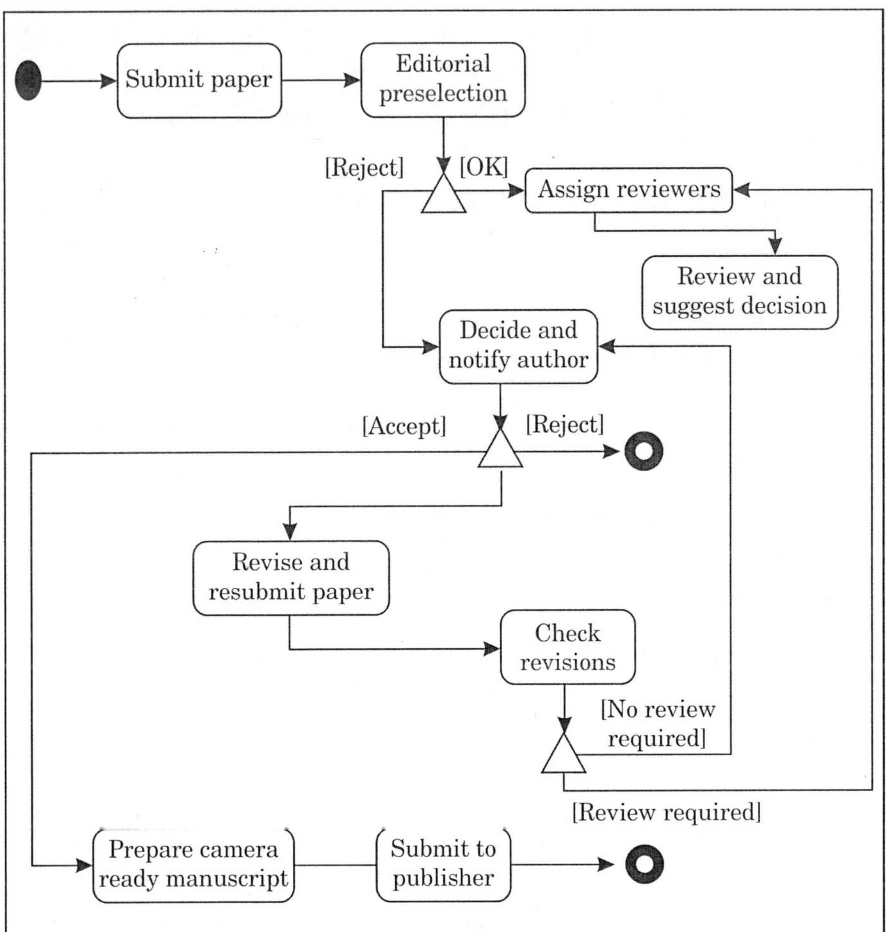

Fig. 3.2: Process of Publication of a Scientific Paper (Michael D, International Journal of Technology and Enhanced Learning, 2014, 1-20).

As a common process of reviewing, any paper received is first primarily scrutinised by the editorial board and then sent to the reviewers for single, double or triple review. Mostly the double blind review is carried out. After getting the recommendations from the reviewers the paper(s) is rejected/selected or sent to the authors for suggested improvements and after the two or three rounds if satisfactory improvements are made by the authors and other conditions are fulfilled like, submission of copyright form, declaration of conflict of interests, the paper is finally selected for publication. A flow chart is shown to describe the publication process.

COMMUNICATION WITH EDITORS

Normally the paper is submitted to the editor of the journal. It is necessary that a paper should always be communicated with a covering letter. It should necessarily be mentioned that the paper is not submitted elsewhere for publication and is based on the original work performed by the authors. Sometimes as per the need of the journal, it is necessary to disclose conflict of interest and acknowledgments in the covering letter itself. There should not be any special requests to the editor to consider the paper for publications on priority or something.

HANDLING REFEREE'S COMMENTS

After submission of the paper, the paper undergoes to the peer review process. The assigned reviewers then judge the paper on some fixed parameters and if found some deficiencies; the same are communicated to the authors through editor or editorial board of the paper. Reviewers point out deficiencies or suggestions to improve the scientific content. The authors are now required to correct those deficiencies and resubmit the paper if they wish to continue with the publication process. While correcting the papers or handling referees or reviewers comments, the following points should be kept in mind:

- Read their comments carefully before answering them. Try to understand the point of the view of the referee and answer accordingly.
- Do not irritate or blame the reviewer for his/her misunderstanding or over sightedness. The referee can be wrong sometimes so make your point with full justification and evidence.
- One should be very polite while answering the questions or when anyone is disagreeing a reviewer's comment.
- The changes should be made exactly as per the suggestions and point-by-point explanation should be made in the text in response to reviewer's comments.
- The paper should be read carefully once again before resubmitting the paper for consideration.
- Errors in statistical data are very common so one should always recheck the data and statistical models applied should be mentioned specifically.
- Submit the revised version within the stipulated time.

GALLEY PROOFS

Once the copy-editing is finalised *i.e.*, the manuscript is checked by the editor to improve the formatting, style and accuracy of the text, the written material or the manuscript is then set into type or the data are processed for printing (typeset). This process gives the proof of the manuscript known as "galley proof" (sometimes called "galleys"). This is then sent to the author for checking the accuracy of the manuscript before it goes for final printing, as errors and omissions are likely to occur at the typesetting stage also. If any serious printing errors exist, even though the manuscript is perfect, it can totally destroy comprehension. Moreover, it may even ruin the reputation of the author as a researcher. Even the small or minor error, like a decimal point is misplaced, can lead the scientific integrity of the paper in doubts.

Hence, it becomes the author's accountability or liability to carefully read and check the copy- edited proof of the manuscript. Sometimes the editor posts some queries also which should be carefully and properly answered. For all this process generally a stipulated time (usually a week) is given to avoid any delays in the publication of the manuscript. However, if the author is under the impression that the changes asked by the editor may distort the script from its genuine explication, he should point out these at this juncture. To ascertain that the structure and flow of the manuscript is clear, it may still be returned to the author for minor rewording, re-organizing or rewriting of certain portions of the text.

While going through and checking their galley proof, the author is presumed to check any typographical or errors of omission or deletion. Utmost care must be taken while checking the data provided in the tables etc. Authors are required to go through each figure or numbers along with their decimal point carefully as in typesetting the errors in tabular materials occur often. And it is easy for the author to detect these errors as they know the data more than anybody else. It is not so simple for the proof reader or copy editor to identify whether a "15" should actually be "0.15" or "1.5", or whether it placed in the right column or not.

Also the authors should carefully scrutinise the figures, sketches or the illustrations given in the proof as it will absolutely appear in the same way in the final print as shown in the proof. Authors are also required to check that they are replicated effectively, with adequate resizing, appropriate orientation, and sufficient variance (contrast) and intensity (sharpness). Last but not the least; the authors should check their names and coalitions or associations because it is the most basic part of information of the paper. Yet, it is commonly taken for granted and generally overlooked by authors.

Rectification of the Proofs

To ensure that the paper is accurate and that the flow of information is comprehensible, it is advisable to read proofs at least twice. While examining the proofs, the author should mark corrections and answer to queries on the proofs. Several ways of marking corrections are available. One best way is to mark them twice; at the place in the text where the error exists and other in the margin opposite to it. Corrections made should be clear and legible. One should use an ink of a different colour to make it easily visible.

Margin or proofreaders' marks are used to identify errors. These marks are globally used and understood by all publishers. Corrections should be specified neatly and clearly so that requisite corrections are made properly. It is important to understand that, at the proof reading stage of paper publication, no major or considerable revision, rephrasing, rewriting, addition of new tables or figures, or any other significant changes can be made. Usually, the editorial office does not entertain such major changes in the text or illustrations at this stage.

After making corrections, authors now send it back to the publisher for printing. The designer now inputs all the corrections into the layout and produce "page proofs", which is almost the final version, for editing and checking purposes. At this stage, almost all the corrections have been made and the paper is ready for the printing version. This final version is exactly what will be visible on the pages of the printed (or online) journal after its publication.

4

Writing a Review Article

A review article is a complete, explicative scrutiny of the literature in a specific field through summary, classification, analysis, comparison. It is a scientific text relying on previously published literature or data. New data from the author's experiments are not presented (with exceptions: some reviews contain new data).

A review article is written to organize and evaluate literature, identify patterns and trends in the literature, to synthesize literature and to recognize the gaps in the investigation, experimentation, analysis, examination, research and commend new sectors of research Review articles are meant for oracle in distinct research areas, undergraduate, postgraduate, scholar or novice researchers and the persons who are the conclusion or decision-makers. Review articles targeted at the last two groups: Extended explanations of subjects or of subject-specific language are mandatory (*e.g.*, through the uses of information boxes or glossaries).

TYPES OF REVIEW ARTICLES

Based on the Methods Adopted

Narrative review

Based on the author's experience, previous theories and models, the chosen studies or surveys are compared and the results are summarized. Conclusions are mainly depicted on the basis of quality rather than quantity.

Best evidence review

A main focus is on the combination of selected studies or surveys with standard methods of study-selection and result examinations.

Systematic review

The researcher's conclusions from several individual studies are examined statistically through stern procedures. Then, the results of the above are put together by the use of Meta-Analyses.

By Objective

Status quo review

It is based on presenting the most recent research on a given subject or an area of research.

History review

It refers to the evolution of an area of particular research or study over a period of time.

Issue review

It involves the thorough evaluation of an issue (*i.e.*, a point of argument or controversy) in a particular area of research.

Theory/model review

It is based on the incorporation of new theory or model in a specific area of research.

By Mandate

Invited reviews

Proficient or competent researchers are invited

Commissioned reviews

Official contracts between the authors and the clients

Unsolicited submissions

A new proposal for a review is thrived by the researchers and is submitted to journal editors.

Flow chart depicting the types of review articles is shown in Fig. 4.1.

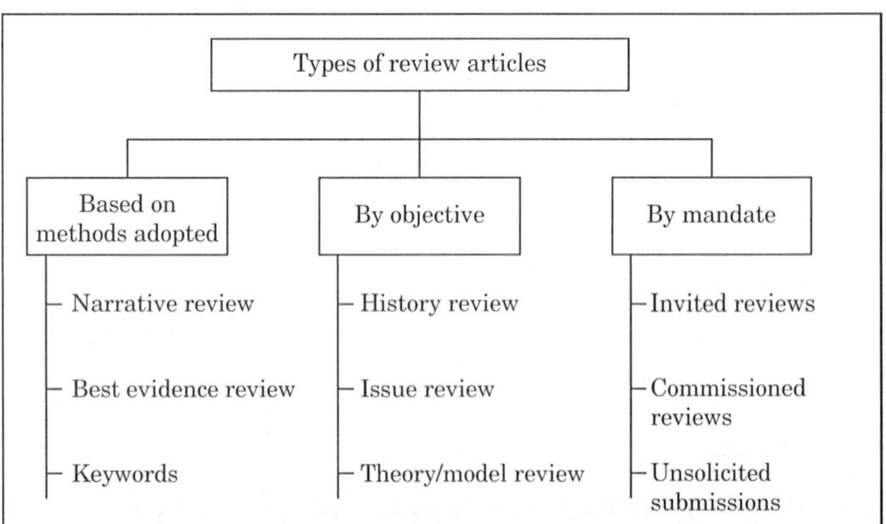

Fig. 4.1: Flow chart depicting the types of review articles

LENGTH OF A REVIEW ARTICLE

Review articles vary considerably in length. Systematic reviews are generally shorter having less than 10,000 words whereas, Narrative reviews may have words ranging between 8,000 and 40,000 (references and everything else included).

ELEMENTS OF A REVIEW ARTICLE

Title

It helps the readers to decide whether they should read the text or not. It includes terms for indexing (*e.g.*, in data bases). The title must be informative and must include important terms. It has to indicate that the text is a review article. It may include the message of the article, not just its coverage. The title must be short and concise but the longer subtitle may be an option only in case a specification is necessary. The present tense stresses the general validity of the results and illustrates what the author is trying to achieve with the article; the past tense indicates that results are not established knowledge yet. It should have length between eight to 12 words. The title may be a question but only if this question remains unanswered at the time of writing.

List of Authors

One should declare intellectual ownership of the work, provide contact information. Every person that contributed significantly to the literature search, literature exploration and/or writing process. The first author should have done most of the research and written major parts of the article. Authors, between first and last author who have contributed in one way or the other to the success of the project, may be ordered alphabetically (indicating equality) or in a sequence of decreasing involvement. The last author usually have coordinated the project and had the original idea.

Abstract

Abstract informs about the main objectives and result of the review article (informative abstract) or indicates the text structure (descriptive abstract). Description of subjects is covered without specific details in an abstract. A descriptive abstract is like a table of contents in paragraph form. The abstract should always be written in present tense. The abstract may be structured as follows:

1. ***Objectives:*** One or two sentences describe the context and intention of the review.
2. ***Materials and methods:*** One or a few sentences provide a general picture of the methodological approach.
3. ***Results:*** A few sentences describe main outcomes.
4. ***Conclusions:*** One or two sentences present the conclusion (which is linked to the objectives).

The length of an abstract should be of 200-250 words.

Table of Contents

It shows the readers the organisation of the text. It helps the orientation among sections. Some review journals print an outline/ table of contents at the beginning of the article, others do not. In general, these are recommended for extensive narrative reviews.

Introduction

This section provides information about the context, indicates the motivation for the review, defines the focus, the research question and explains the text structure. The elements of introduction are subdivided below:

Subject background

This gives the information regarding the context by describing the general topic, issue, or area of concern.

Problem

Current areas of interest, new aspects, gaps, controversies, or a single problem is stipulated or specified.

Motivation/justification

This specifies the reason for author's reviewing the particular literature; the approach adopted and the organization of the text.

The introduction section should be written in present tense (use past tense for the description of your methods and your results). It contains many citations and has length between 10% and 20% of the core text (introduction, body, conclusions). Make sure to have a narrow focus and an explicit research question. Indicate these two points clearly in the introduction. Give theoretical or practical justifications of the need for a review.

Material and Methods

Systematic and best evidence reviews have a methods section. This section enables motivated researches to repeat the review. Narrative reviews do not have a methods section but should include some information about applied methods at the end of the introduction. The materials and methods section contains for example an information about: data sources (*e.g.*, bibliographic databases), search terms and search strategies, selection criteria (inclusion/exclusion of studies), the number of studies screened and the number of studies included, statistical methods of meta analysis. This section is written in past tense and has few citations *e.g.*, statistical analyses or software used. It has length of approx. 5% of the core text (introduction, body, conclusions). Make sure that data sources are clearly identified. Precision has first priority in the material and methods section.

Main Part of the Review Article

A coherent structuring of the topic is necessary to develop the section structure. Subheadings reflect the organisation of the topic and indicate the content of the various sections. Possible criteria for structuring the topic are:

- Methodological approaches
- Models or theories
- Extent of support for a given thesis
- Studies that agree with another *versus* studies that disagree
- Chronological order
- Geographical location
- Avoid referring to only one study per paragraph; consider several studies per paragraph instead.
- Frequently link the discussed research findings to the research question stated in the introduction. These links create the thread of coherence in your review article.
- Link the studies to one another. Compare and discuss these relationships.
- While writing this section, three tenses are frequently used:
    - *Present*: reporting what another author thinks, believes, writes, reporting current
- Knowledge or information of general validity, *e.g.*, It is believed...

  - *Simple past*: referring to what a specific researcher did or found, referring to a single study, *e.g.*, They found...
  - *Present perfect*: referring to an area of research with a number of independent researchers involved, *e.g.*, They have found...

- Citations in this section are usually indirect but in some cases pointed and relevant remarks might be cited directly. Like,
- ***Non-integral references (indirect)*:** The author's name, or a number referring to the reference list, appears in brackets. Non-integral references emphasize the idea, result, theory etc. rather than the person behind it. Most references in biology are non-integral.
- ***Integral references (direct)*:** The author's name has a grammatical function in the text. As Ridley (2008) points out this type is appropriate to emphasize the contribution of a specific author.

This section covers 70 to 90% of the core text (introduction, body, conclusions). Make sure to organize the different pieces of information into a line of argument. An appropriate organisation of information is all-important for the quality of a review. Throughout it is important that the idea/topic (paragraph 3 of the Introduction) drives the article and not the literature used; write an idea-driven, rather than literature-driven article!

Conclusions

It answers the research question set in the introduction. It has elements like, implications of the findings, interpretations by the authors (kept separate from factual information) and identification of unresolved questions. The conclusion is written in present tense. It has few or none citations and covers of about 5 to 10% of the core text (introduction, body, conclusions). Make sure to have a clear take home message that integrates the points discussed in the review. Make sure your conclusions are not simply a repeat of the abstract.

Acknowledgements

This section is to express gratitude to people who have helped in the literature survey, the organizing of the material or in the writing process (but whose contribution is too small to justify as co-authorship). Express gratitude to funding organization and specifies the funding program (often required by funding agencies). In this, full names of people and their specific contributions to the project are given. The name of the funding agency and program as well as the grant number and the person to whom it was awarded are mentioned. This section should be written in present tense (past tense when referring to funding agencies in terminated projects). It contains no citations.

References

References show interested readers how to find the literature mentioned in the text, acknowledge the work of other scientists and is compulsory to avoid charges of plagiarism. It includes every reference cited in the text. Do not include additional references. Avoid internet sources. If internet sources must be used, find the original source for the internet reference, check it has been correctly cited and cite it directly. Length of this section ranges between 50-100 references in most cases appropriate. For narrative reviews the inclusion of all relevant, high-quality studies is the target. Moreover, systematic and best evidence reviews need explicit criteria for the inclusion/exclusion of studies from which they got the data.

Illustrations

Illustrations or concept maps are used in review articles to visualize the structuring of the topic, to show the relationships between studies, concepts, models or theories. Organisation of data boxes with terms or names are arranged in a two-dimensional space. Arrows are used to link boxes. Specifications of the relationship are written on the arrows. The legend describes the concept map's content. It is specific and informative (*i.e.*, it should be possible to understand the map without reading the full text). Concept maps are very useful to display complex relationships.

PREPARING A REVIEW ARTICLE IN 18 STEPS

1. Narrow the topic, define a few research questions or hypotheses
2. Search for literature sources, refine topic and research questions during the search
3. Read, evaluate, classify and make notes
4. Redefine the focus and the research questions, define the take-home message
5. Compose a preliminary title develop structure
6. Find a structuring principle for the article
7. (*e.g.*, chronological, subject matter, experimental procedure)
8. Prepare an outline, find headings for the sections in the text body
9. Plan the content of each paragraph in the different sections
10. Prepare tables, concept maps, figures write draft
11. Draft the methods section (if needed)
12. Draft the body sections
13. Draft the conclusions
14. Draft the introduction

15. Draft the abstract revise
16. Devise drafts of different sections, abstract & title, tables, figures & legends
17. Revise citations and references
18. Correct grammar, spelling, punctuation
19. Adjust the layout

5

Delivering a Scientific Oral Presentation

ORAL PRESENTATION

It is a type of scientific communication in which speaker deliver his message related to any research topic in a clear, organized and stimulating fashion with the help of presentation. Before writing a presentation a speaker should think about the audience, expectation from the speech and how much time will be given to deliver the message to audience. For preparing oral presentation there are several steps which should be followed.

1. Preliminary steps for preparing an oral presentation
2. Organizing and writing the speech
3. Practicing the speech and handling logistics
4. Phrasing the speech
5. Managing stage fright
6. Visual aids

PRELIMINARY STEPS FOR PREPARING AN ORAL PRESENTATION

There are several preliminary steps for preparing an effective oral presentation.

Analyzing the Audience

Before deciding a topic for oral presentation, it is important to figure out certain things as how to make one's presentation interesting, time you can devote to it, how much audience will be there to listen to you and how loud you will have to be so that the complete audience can hear you. All these facts will help you to ascertain the depth of your discussion, the audio and visual assistance that you can use, and the conditions or the circumstances for your presentation.

Selection of a Topic

The topic chosen should be such that it thrills you and in which either you have the expertise or which inspires you to have one. At time you may not have the choice to select the topic of discussion, but still irrespective of the subject, you can plan or chose the modus operandi. The topic should be such that it shows some correlation with the current issues prevailing in the society. Timeliness can make a presentation more interesting to your audience. In order to focus on a topic follow the following steps:

1. Determine your general goal
2. Develop a precise objective
3. Develop a title
4. Develop a summary

Researching the Topic

Accumulate as much as information regarding the topic chosen than you feels to be required. Compile the information collected in the same manner as it should be done while writing the research paper. Look through your accomplished notes and identify each component or segment with a particular number or word that notifies you about, where each thought fit into your outline.

ORGANIZING THE SPEECH

Oral presentation is different from the written information. Unlike published paper, in oral presentations the audience can't turn back and examine the first part of your speech. That's why the speakers generally repeat themselves while giving the presentation. The simple technique for delivering the speech is "First, acquaint the audience with the details of what you are going to speak about; then tell them; finally recall all that you have told.

Before writing your speech, first prepare the abstract or the summary of your speech. Write down the main points of your speech, and then start elaborating them one by one, giving different headings and subheadings. This will finally help you to frame the complete speech. Most word processing software includes an outlining feature, which may help.

Besides the main points, the speech should also have an introduction as well as a conclusion. The introduction part of your speech should be such that it attracts the audience's attention and warm you up.

PRACTICE OF THE SPEECH

Practice delivering your speech before presenting it. Moreover, if practiced in a simulated environment as to where you'll be giving the presentation, it will be more effective. If possible, practice in front of people, your friends etc. and use your visual aids. Your speech should be a blend or combination of information, entertainment, and intellectual stimulation; all expressed in a spontaneous and comfortable manner. The following guidelines should be kept in mind.

1. Do not memorize the script and deliver it exactly the same.
2. Write in large, boldface letters
3. Try to record yourself and listen to the pitch, tone, and speed of your voice
4. Practice your speech out loud
5. Pause naturally as you would in conversation.

Logistics

Assure that all the appliances, hardware/software or any aid you are using for the presentation is in working condition in advance. Analyze the problems that might crop/turn up during the presentation, and find the ways to tackle with them. Like, "what will you do if the previous speaker takes longer time and, and you have been asked to condense or cut short your Speech?", or "what will you do if your presentation is deleted or the computer crashes?" Hence always be prepared for all the fallacious things that can happen.

If the venue for your presentation is not familiar to you, check out the space on the eve of your presentation. This will help you to know the details of the area of that place and to get an idea of what technical aids you can use. Moreover, it will give you an overview of all the facilities available there like availability of internet hook-up, projection screen and their working etc. These all details will help you to deliver an effective speech.

PHRASING THE SPEECH

During oral presentation different languages can be used but not in research paper. A speech need not be sound as report. Audience will grab only when you speak in a language which is familiar to them. Oral communication is different from written in that the oral communication is

1. Instantaneous
2. More interactive
3. More audience-specific
4. More personalized
5. More informal or friendly
6. Gives more opportunities to use visual communication

Oral presentation not only differs in language but also in style from a research/review paper. Here are following consideration for phrasing oral presentation

1. Use vocabulary that will be understood.
2. Use enumeration to tie your points together.
3. Use parallel construction in your phrasing to help the audience follow what you're speaking.
4. Use personal pronouns and refer to yourself and the audience.
5. Interject ideas and comments.
6. Ask occasional questions.

MANAGING STAGE FEAR

It is an important part of oral presentation. Speech can become ineffective due to stage fear. Many people thinks that stage fright is caused mostly by what happens during the speech but in actual it happens mostly due to state of mind before the speech and sometimes due to the speech itself. Overcoming of stage fright is necessary and can be overcome by following way.

1. Practice in place or situation similar as possible.
2. Watch or listen other speeches, either audio or video
3. Use mental image to picture yourself in front of an audience.
4. Try to become comfortable with an idea.
5. Perform voice and breathing exercise before presentation.
6. Try to establish dialogue with the audience.
7. Take pause and breathe normally during speech.
8. Have water with you during presentation.

VISUAL AIDS

Visual aids can be used during oral presentation to make it effective but there should be some points which must be keep in mind during oral presentation

1. Use colors but not too much.
2. Break complex ideas into simple visual parts.
3. Keep visual aids simple.
4. Direct your presentation towards audience instead of talking at your visual aid.

There are several options which can be used as visual aid with their advantages and disadvantages.

Overheads

Overheads are simple and clear, and you need not depend on a computer. They can, however, have poor print quality, get non functional, and cause other problems if the transition between each one is not smooth. If possible, tell someone else to turn your overheads during your presentation, which will enable you to concentrate on speaking and directing the complete presentation.

PowerPoint or Similar Slide-show Software Programs

This gives a professional-look to your presentation. A typical oral presentation is as shown below in Fig. 5.1. You can save your presentation on a CD, pen drive etc and carry it with you. Moreover, you can make any changes, deletion or insertion of any information into your presentation easily. Besides having the advantages this makes you dependent on the computer technology, and if there is any technology problem, or, if the computer crashes, you won't have your slide-show. Therefore, whenever you are using such software, be prepared with alternate visuals such as overheads or paper handouts.

Fig. 5.1: A typical oral presentation session

Slides

Slides give clear images of all the photographs etc. It also makes easy to change the order of the slides if required in your presentation. However, they are comparatively expensive, and ones the images are created they cannot be changed.

Whiteboards and Paper

This is more convenient method to explain your points, if you are comfortable writing in front of the audience, as it gives you the spontaneous and incorporate feedback from the audience. However, it does not give the professional look as other media, and it takes most of your time writing (often with your back to the audience) when you should be talking.

Handouts

Handouts, although a peerless accompaniment to any of the options listed above, they themselves too have their own disadvantages. If the handouts are distributed to the audience at the beginning of the presentation, it may lead to the risk of losing audience's attention, interest etc. as the audience will be busy reading the handouts rather than listening to your presentation.

In addition to above there some common tips which can make oral presentation effective

1. Warm up with breathing and vocal exercises.
2. Take a deep breath before walking to dice/ stage.
3. Take a pause and look at the audience before starting your speech.
4. Be comfortable with the sound of your voice.
5. Remember to pause in between.

6. Completely focus on the audience, besides visual aids or your notes.
7. Change the pitch of your voice giving stress on some particular words or sentences that are important. This will draw the audience's attention and will also make your presentation attractive and effective.

After finishing, invite the audience for asking the queries and also try to get the feedback. Moreover, always remember that the number of times you go through this process the more efficient and good public speaker you will become.

6

Delivering a Scientific Poster Presentation

POSTER PRESENTATION

Amalgamation or combination of verbal presentation with visual aid leads to a perfect communication tool represented as a scientific poster. Poster presentation is one of the conventional methods to represent results of any program assessment, statistical analysis, or any other project at the professional conferences. And a scientific poster is method to represent or display the results of any research or scientific project at a professional conference.

Developing a poster does not include only fabricating the pages to be displayed in the conference, but also preparing the necessary handouts, and preparing the expected questionnaire and queries likely to be confronted during the session. In general, poster presentation gives more detailed information than a speech but less than a written paper. Moreover, it is also more interactive than either of them. The major difference in oral presentation and poster presentation is that in oral presentation the speaker or the presenter establish the focus of the presentation, but in a poster presentation, the viewers drives that focus. Different people will have different queries about your research. Posters made should be such that it has the same professional touch as your research. The following guidelines will help you to prepare an effective and successful poster:

1. Poster preplanning and preparation
2. Poster assembly
3. Poster session

PREPLANNING

Before selecting the topic of poster presentation one should keep following guidelines in mind.

Writing for a Varied Audience and Hence Make Poster Language Simple

Audience, viewers or listeners at professional conferences differ considerably in their substantive and methodological backgrounds. Some will have expertise in the topic/ or the area of your research, some in the analytical techniques/technologies used by you in your research methods. Always the poster presentation should be in simple language which can be easily understood by all type of the audience clarifying your object of doing a particular research, your expected outcomes and the benefits you can gain through that research in few minutes by giving a glance at your poster.

Abstract/ Summary

It should be informative and brief.

The Title

Title should be such that it draws the attention of the viewers. It must be written in large, clear and bold (preferably solid-block) letters to be easily read from a distance of at least 5m. Use either single or not more than two font styles; else it will cause distraction while reading. Always try not to use upper case letters; it will also make reading difficult. Even you can introduce the institution or organization logo to whom you are representing.

Text

Unlike research paper, in the poster presentation, more information is given in a concise manner. The area allotted for poster is definite and you have to display all you work in that confined area. Present your findings as main points using bullets and numbers in the required font size which can be explained verbally whenever asked. Before the final print out is taken just take a look at your poster, revise it, check the spellings etc.

Authors

While writing the author's name in the poster, only include first name and academic qualification. Omit middle initials, city and titles like Mr., Dr. Etc.

Explaining Statistical Methods

While writing about statistical methods, relate them to the specific concepts of your study. Give synonyms for technical and statistical terms used because people from a wide range of disciplines come to attend the conferences etc.

Presenting Results with Charts

The most preferred way to depict numeric patterns, relative sizes of groups, comparative levels of some outcome etc. is presenting them in tabular form, line graphs, bar charts or histograms etc. Do not forget to label each chart with a suitable heading that describes the topic of that particular chart.

Photos and Visuals

Visuals can be used during poster presentation to make it effective but there should be some points which must be kept in mind during poster presentation.

1. Do not use many colors.
2. Break complex ideas into simple visual parts.
3. Keep visuals simple.
4. Do not talk at your visuals.
5. Photos should be of an appropriate size according to the poster.
6. Use colour photocopies when required for quick, cheap, good quality reproductions.

Content

The content or matter to be presented as poster should be in a very concise manner. Following points should be kept in mind while presenting the poster.

1. Poster lay-out should just focus on the main points.
2. Outline the expected queries and describe the expected applications of your research work.
3. Try to express complex statistical procedures/methods in a simple descriptive manner.
4. Acronyms if used should be elaborated.
5. Express the large detailed results/tables in the form of charts/ models /flow charts/graphs etc.

Handouts

Like oral representation, handouts also have the similar

disadvantage in the poster presentation. They lead to the risk of losing audience's attention. Handouts should only include abstract/ summary of the research work carried out and contact information. Moreover, it should be handed over only to the interested audience.

POSTER CONSTRUCTION

Scientific poster presentation is analogous to writing a scientific paper, with distinct pages dedicated to individual topic like background and aim of the project, materials and methods, tables and graphs, results and discussions, and conclusions. The preliminary introduction should include the small description about the topic, its importance, the expected results and its benefits to the society. The material and method part should describe in brief the methods adopted to analyze, how the data were gathered, analyzed etc. Depict the results or the outcomes of the research project in a simple and expressive manner. In the conclusion part, discuss the expected outcome of your work and its use in the society. After designing the paper layout, determine the available space assigned to you. Finally display them in such a way that it makes the viewers to read easily from left-to-right and top-to-bottom. Here are some guidelines for poster construction.

POSTER LAYOUT

1. To begin with, first at the top place the title of your work in bold and of readable size. You can also keep some space for your organization's logo.
2. To present an overview of your research work, assign one sheet for the brief abstract or summary of your work. It should be written in such a manner that it is self explainable and enables the reader to decide his area of interest.

3. On the left-hand side, give the introduction about your project, its importance, its need, work carried out on the related subject, your main aim for carrying out the study and also how your work is better than the previous studies or analysis.
4. In the middle, describe the materials and method adopted, data obtained and the findings or results represented in the tabular form or charts associated or supplemented with flow charts, diagrams, photographs, text etc.
5. Then, on the right-hand side, give the brief discussion of your findings relating it to your aim of study, its advantages and disadvantages, limitations of your work, future prospects and the possible benefits to the society. Standard poster layouts are as shown below in Figs. 6.1, 6.2, and 6.3.

Transporting the Poster

Laminated poster are easy to display and does not require any tedious job but, they need to be protected while carrying as their size may be large and require a poster tube to shield it. Always have an identification mark on your poster so that it may be easily detected if lost.

Poster made on loose separate sheets or cards of reasonable size are comparatively easy to carry but it has its own disadvantages. You may lose any sheet during transportation or while pasting and even you have to keep in mind the sequence of the sheets if not numbered.

It is always better to make a duplicate poster and keep so that if required it can be easily available.

POSTER SESSION

During the poster session, the author is required to stand by the poster to reply or answer the questions or queries of the viewers.

Always stand at the side in order to provide the clear view of your poster while explaining to a particular spectator. Don't ignore people who appear to have some queries. Follow instructions pertaining to poster removal allowing the space for the next presenter in another session. You can also ask somebody to take your photograph with your poster as proof of your scientific presentation.

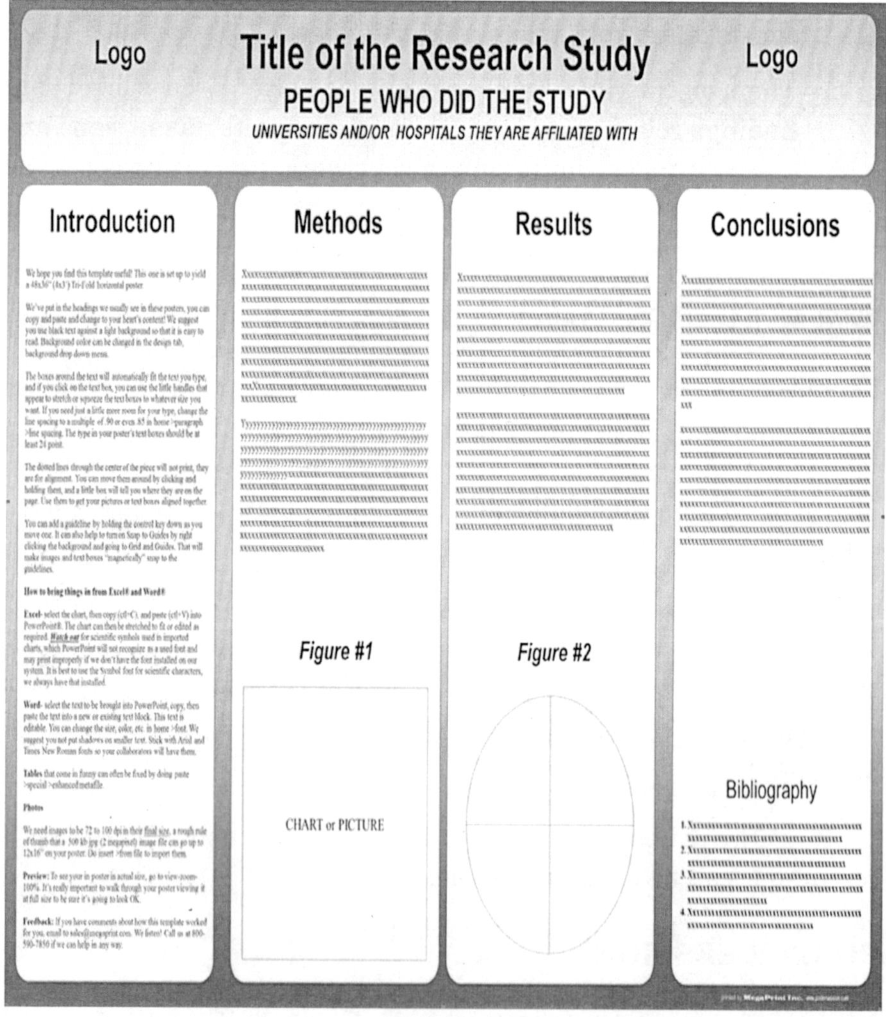

Fig. 6.1: Standard poster layout

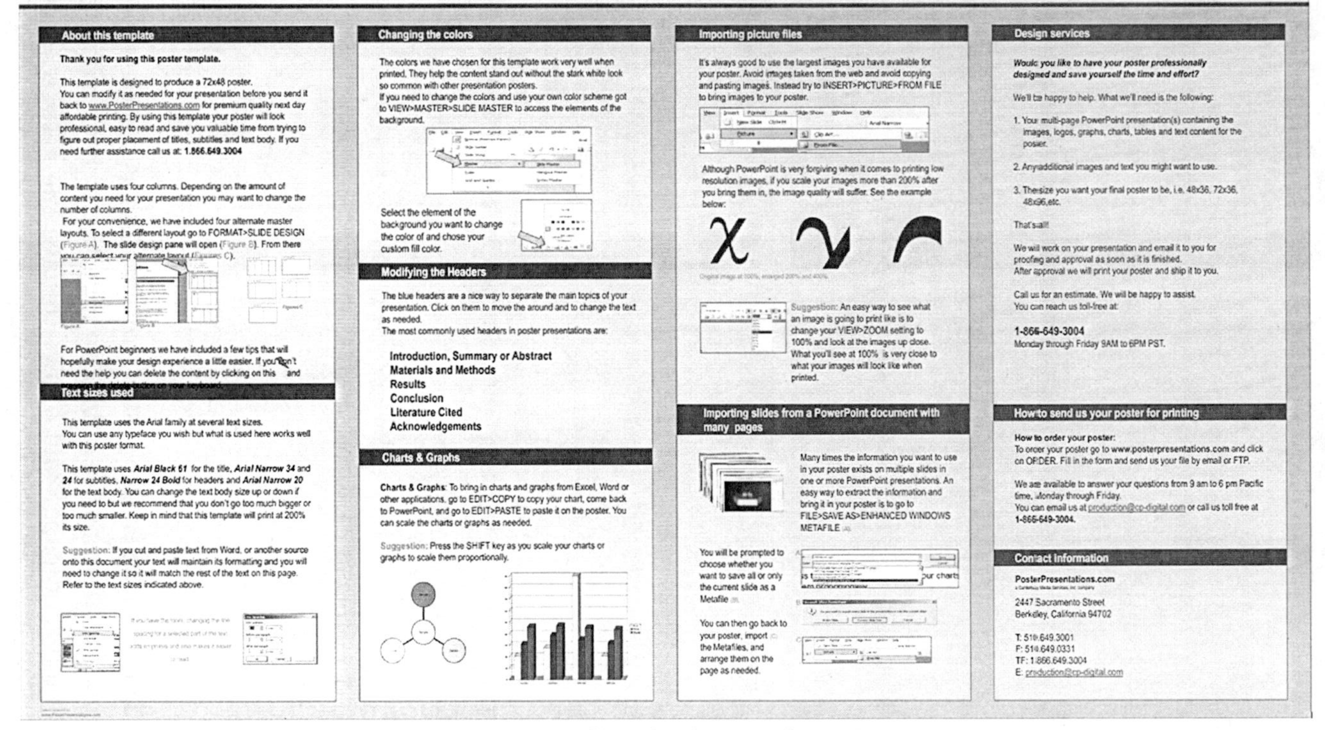

Fig. 6.2: Standard poster layout

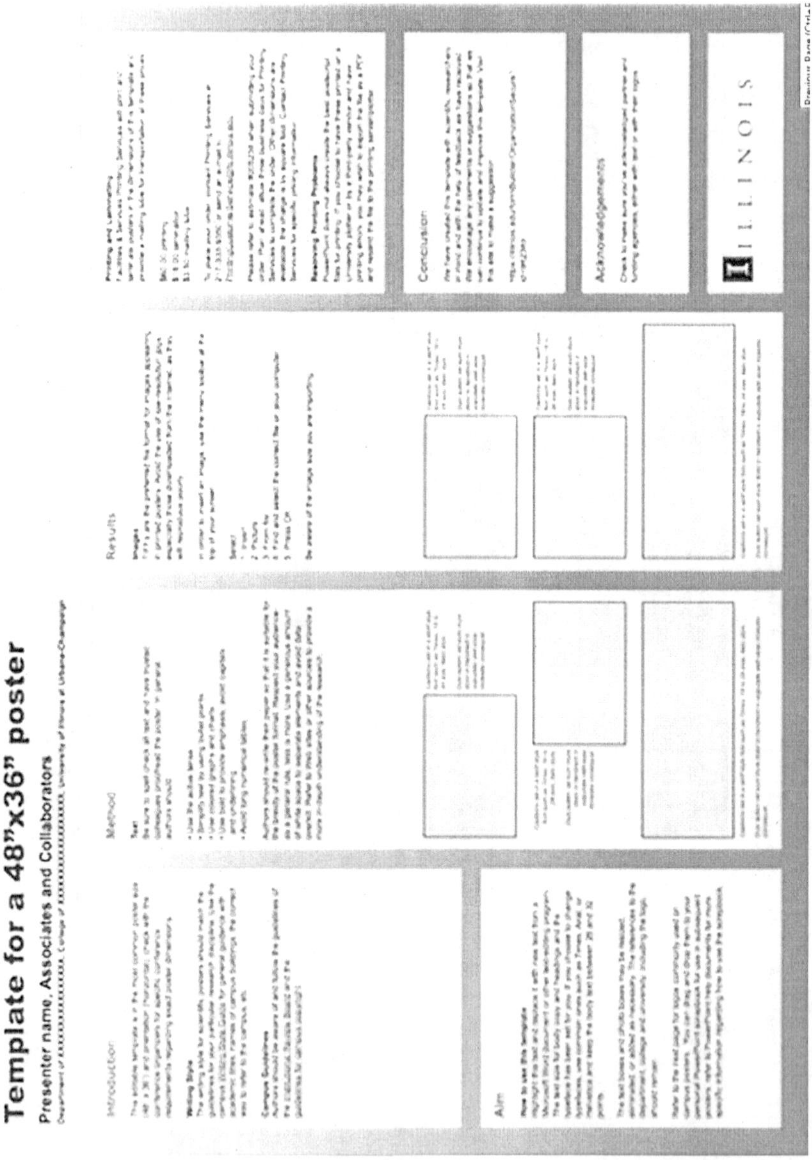

Fig. 6.3: Standard poster layout

On the whole, the poster presentation is a two-way means of communication, as it enables to give the written information like the published paper and allows to have the detailed conversation with viewers as in case of oral presentations.

7

Writing Practical Reports

INTRODUCTION

Practical report is preparing a report on an experimental procedure carried out in a lab. It is a routine but very essential part of scientific studies. While writing a report the main focus should be on giving the complete facts and accurate information regarding your investigation or study. It should give the complete account of your work *i.e.*, what has been done, your findings, and the discussion and conclusion of your study.

The writing of laboratory reports plays an important role in practical course as it helps you to understand how to execute while writing such reports.

The main motive of writing practical reports is to convey the readers the idea behind the objective of doing that specific work, methods adopted, expected outcomes or the results and the impact or the benefits of the study to the society. Some readers, instead of reading the complete report, look for some precise information for their queries. It is because of this, it becomes a necessity that one follows a standard format, which includes, headings, subheadings etc. This enables the reader to find the information required immediately without going through the entire document.

The most simplified rule for writing the report is '*whether the information provided in the report could be replicated or not*?' It should be kept in mind that the reports are generally written with the concept that they are read by those persons who are ignorant about your experiment. Hence, the first thing they will read about your report will be the abstract. And after that only they will go through the entire report.

No such particular format is available which is designated to be more 'correct, or perfect' than any other. However, some widely adopted standards and conventions are followed. If you are not sure about the correct format or style, just browse the British Journal of Psychology.

SECTIONS OF A PRACTICAL REPORT

A practical report consists of various sections describing the particular or specific details. The main sections of the report are:

1. Title
2. Abstract/ Summary
3. Introduction
4. Materials and Methods
    - Materials and instrument / apparatus used
    - Procedures
5. Results and Discussion
6. Conclusion
7. References
8. Appendices

Title

The title of the report should be such that it is clear, concise and reflects the work done. Remember that your title only will help the reader to identify whether the report is of his/her area of research or not and that he should go for complete reading of the report or not. Hence, the title should be a true imitation of your work in brief.

Title should be apparently between 10 and 20 words but not more than that. It can have beyond 20 words only in exceptional circumstances.

Abstract

The abstract gives the self explanatory brief summary of the detailed write-up. It helps the reader to identify and validate whether the write-up is of his /her interest or not. Basically, the abstract includes the brief description of the issue/problem or the topic of write-up, the methods used to evaluate or investigate the matter and the main results/ outcomes derived. It may extend up to approximately 100 words.

The abstract is comparatively easier to write if written at last after completing the entire document although it comes at the beginning of the write-up.

Introduction

It gives the description regarding the study which is undertaken for research, the objective of doing the particular study, its background or the brief history, previous studies done so far, new approach targeted, methods that will be adopted to achieve the required results and finally the expected results and outcomes. In other words introduction intends to summarize the background and reasons for carrying out such a research. It should also describe the significance and current relevance of the study. For this, relevant matter from journals, textbooks and other resources can be taken and cited which should be referenced properly.

While writing the introduction following information should be lined-up as below:

- A brief review of the previous studies done on the subject selected
- Brief description of the problem undertaken for study and the procedures adopted to investigate it.
- Finally, the description of the predicted results according to your experimental hypothesis (hypotheses)

In short, introduction is mainly described in two parts: the past research, and new targets of the study. The previous work or the past studies helps to understand the phenomena that are investigated in your study and illustrate the limitations or drawbacks of the previous work. It may be related to methodological problems, or it may have further scope for extending the work or it may also require replicating the results in a different manner or methods. If the previous work is up to the mark, then there would be no scope for the further study. The second part of the introduction should be focused on the actual work done in the present study.

Materials and Methods

This section of the report gives the detailed information regarding how the work was done, using what means and methods. This section is of great importance keeping in view the methods used as this can be treated as standard for further experiments. There should be a proper balance between insufficient information and an over-detailed description of the events. You should keep in mind that your work should be reproducible *i.e.*, if repeated by someone; he/she should get the similar results or conclusions.

This section should be written as a detailed description of what is done and how it is done.

Materials and methods

This section gives the information regarding the apparatus and materials used in the experiment. If some special or typical instrument or equipment is used, it should be explained in detail, also use a diagram, if required.

This section gives the detailed description about the procedure adopted to carry the experiments. The information provided should be self sufficient so that if required the experiment can be reproducible.

The statistical tests used should not be included in this section, as they are only used for analyzing data and not collecting the data.

RESULTS AND DISCUSSION

In this section the writer should give the complete, clear summation of the data accumulated or gathered with the results and discussions. All the tables and figures should be well labelled that they are self-explanatory. It should not mean, that they need not be explained in the text but, instead they may not cause the inconvenience to the reader to understand them without its help. Some explanatory text must be included describing data provided in the tables and figures. Some simple do's and don'ts for representing the data are:

DO'S	*DON'TS*
Tables should be presented in appropriate measures of central tendency (mean, median or mode) and dispersion (variance).	Don't discuss about "differences" in the data unless it is analyzed.

Table (*Contd...*)

Table *(Contd...)*

DO'S	*DON'TS*
If required give the pictorial representation of the results like graphs etc.	Don't describe the rationale behind the statistical test.
Give the details of statistical analysis done for your data.	Do not discuss about "differences" in the data if the inferential test was not significant.
If an inferential test is carried out, give the value of the test statistic, degrees of freedom, observed p-value and state whether the test is a one or two tailed test.	
State whether the null hypothesis was rejected or not.	

In the discussion part, the interpretation of the results of the experiment is described. The interpretation of the results should be such that they reflect the relation to the problem or the issue discussed in the introduction as they provide the basis or the idea behind the carrying out of the experiment. Your discussion should give either the solution for the issue raised or propose any further studies that may be required to be carried out to achieve the expected results. Even the limitations or difficulties encountered while carrying out the experiment may also be discussed in this section which were not initially predicted. This may prove helpful in future experiments. In short, you should clearly depict the questions or the problems taken up by you before carrying out the research and the method adopted by you to solve them. Keep in mind not to just conclude saying future research is required which may leave the reader in confusion as to what should be done next or what the future scopes of the studies.

While writing the discussion part one can start by just summarizing the main aspects of the results and findings. Then discuss the actual meaning of the results and relate them to the problem discussed.

CONCLUSION

Write the final inference you deduced from the results obtained in a concise manner. The limitations, if any, should be presented in such a way that it inspires or propels the researcher or the reader for future studies. This section also discusses how your experiment will benefit the society.

REFERENCES

If any written material (from *e.g.*, texts, journals) is referred for writing report, it should be correctly cited in the report and referenced in an appropriate form here. There are a number of referencing styles available. Be consistent within your style chosen and use an accepted style that is used in science based resources.

E.g. for journal articles

Surname, Initial. (date/year) Title of paper written in lowercase letters. Title of Journal Paper in italics or underlined or abbreviated from, Vol. No: page numbers.

E.g. for books

Surname, Initial. (date/year) Title of book underlined or in italics. Place of publication: Publisher.

Appendices

The appendices may include support material not included in the main body of the report, for example, illustrations, brochures, sets of data etc. This material should not be required to fully understand the report, *i.e.*, the report should be able to stand alone. A journal article does not include appendices.

The appendices should not just be an afterthought and should be as comprehensible as the rest of your study, with clear subtitles for different types of material.

8

Plagiarism Issues in Scientific Contents

PLAGIARISM

Plagiarism is unacknowledged copying the work of a person and presenting it is as one's own work. It does not matter whether the work that is copied has been published or not but main point here is that the work is copied from someone else and is published as one's own work without mentioning the name of the original person.

PROBLEM WITH PLAGIARISM

Plagiarism is considered a major problem in the scientific world because lots of cases are reported. Plagiarism not only infringes upon existing ownership, but also deceives the reader and misrepresents the originality of the current author. If a plagiarism relationship exists between two texts, it suggests that the texts exhibit some degree of same text, manner or style which would not happen if the texts were written independently.

REASONS FOR PLAGIARISM

1. Some students plagiarise unintentionally, when they are not familiar with proper ways of quoting, paraphrasing, citing and referencing.
2. Students plagiarise to get a better grade and to save time.
3. Students' overtaxed lives leave them so vulnerable to the temptations of cheating.
4. Some student see no reason why they should not plagiarise or do it because of social pressure, because it makes them feel good or because they regard short cuts as clever and acceptable
5. To some students plagiarism is a tangible way of showing dissent and expressing a lack of respect for authority.
6. Some students deny to themselves that they are cheating or find ways of legitimising it by passing the blame on to others.
7. It is both easier and more tempting for students to plagiarise as information becomes more accessible on the internet and web search tools make it easier and quicker to find and copy

FORMS OF PLAGIARISM

Plagiarism can take several distinct forms, including the following:

1. **Word-to-Word:** Direct copying of the matter from a published text without quotation or acknowledgement.
2. **Paraphrasing:** When words, manner or style are changed or rewritten, but still the main source text can still be recognized.
3. **Plagiarism of Secondary Sources:** When original sources are referenced or quoted, but obtained from a secondary source text without looking up the original.

4. **Plagiarism of the Form of a Source:** The structure of an argument in a source is copied.
5. **Plagiarism of Ideas:** The reuse of an original thought from a source text without dependence on the words or form of the source.
6. **Plagiarism of Authorship:** The direct case of putting your own name to someone else's work.

Some forms as mentioned above are easily recognized while others like paraphrasing and the reuse of structure are relatively difficult to recognize as the matter is rewritten in such a way that the original is hidden completely. These forms of plagiarism are not just harder to detect, but also harder to prove.

EIGHT ACTS OF PLAGIARISM

(a) Copying text and inserting it in a paper without citation,

(b) Copying an entire paper without citation,

(c) Asking someone to provide them with a paper,

(d) Using the internet to copy text and insert it in a paper without citation,

(e) Using the internet to copy an entire paper without citation,

(f) Using the internet to ask someone to provide them with a paper,

(g) Purchasing a paper from a term paper mill advertised in a print publication,

(h) Purchasing a paper from an online term paper mill.

PLAGIARISM DETECTION

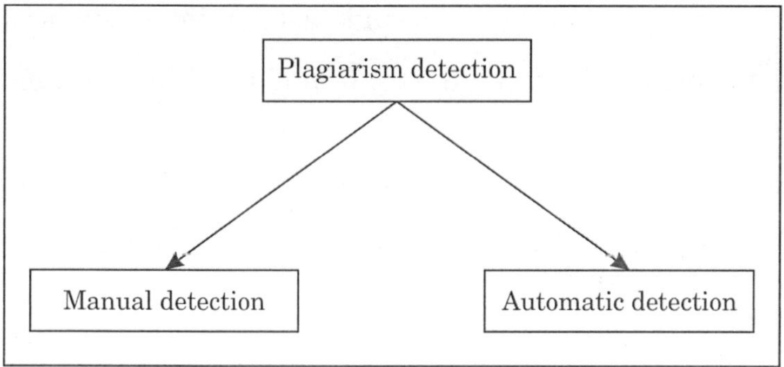

Methods of plagiarism detection

Manual Detection

Manual plagiarism detection in a single text is based on identifying author's writing style, or recognizing passages with a familiar feel to them. Between multiple texts, plagiarism detection involves finding similarities which are more than just coincidence and more likely to be the result of copying or collaboration between multiple authors. In some cases, a single text is first read and certain characteristics found which suggest plagiarism. Ability to detect plagiarism comes with knowledge, experience, interest and time spent by a reader/teacher assessing the work of students. Some of the following points that reflect towards plagiarism to the reader/ checker:

1. Use of advanced or technical vocabulary beyond expectation of the writer.
2. Unexpected improvement in writing style compared to previous submitted work by the writer.
3. Inconsistent text found within the written text.

4. Sudden use of the irrational text in the form of a paragraph where the flow is not consistent or smooth, which may signal that a passage has been cut-and-pasted from somewhere else.
5. Big similarity between the content of two or more submitted texts.
6. Shared spelling mistakes or errors between texts.
7. Dangling references, *e.g.*, a reference appears in the text, but not in the bibliography.
8. Use of inconsistent referencing in the bibliography suggesting cut-and-paste.

Automatic Detection

The main aim of automatic plagiarism detection system is to improve manual detection by, reducing the time spent on finding plagiarism, comparing big or multiple texts and finding their sources. The automatic detection system also provides the ease to the user and even a lay man is able to detect the plagiarism in the given text. Also some errors possible with manual detection like false positives (those incorrectly classed as plagiarised) and false negatives (those incorrectly classed as nonplagiarised), and possibly provides the number of true positives (those correctly classed as plagiarised) and true negatives (those correctly classed as non-plagiarised). But the main problem with the automatic detection is to minimize the false positives and false negatives. Moreover, the results obtained from different automatic detection systems should be similar otherwise the results would become more confusing rather solving a problem.

Software to detect the plagiarism

There are many softwares available in the market which offer automatic plagiarism detection efficiency. Some are free wares and some are paid. Out of them a few are discussed here.

iParadigms has developed a search engine, iThenticate, which is able to clearly identify matching texts between two text based documents of any language or size. The search engine examines a chunk of text, eliminates words that are too common, and turns the other words into numbers. It also converts Internet content into numbers. Consequently, it can compare patterns of numbers from the sample with patterns of numbers from the Internet. To do this, the search engine uses a number of complex mathematical algorithms. iParadigms' search engine is used at Web sites that monitor plagiarism (*e.g.*, Turnitin.com) and at a Web site focused on media such as movies and music (Slysearch.com).

At Turnitin.com, students or instructors submit papers that will be tested against the Internet or proprietary databases for plagiarism violations. The company believes that the recirculation of term papers is a key source of plagiarism; therefore, its database contains student papers, papers posted online, material from academic Web sites, and documents indexed by major search engines. Its database consists of 800 million Internet pages and more than 100,000 papers. Student papers submitted by registered users are also archived to the database. Thus, extended use of the service will build an instructor's archive of papers and will ensure that students cannot easily recycle papers from previous classes. If an instructor is interested in "testing" a paper for plagiarism, he or she submits it for processing through the proprietary search engine. Once the process is complete, an originality report indicates the probability (in terms of a percentage) of whether the paper was plagiarized. Instructors can click on links to direct them to the source of the possibly plagiarized material. This process takes twenty-four hours on average but can take up to two days, depending on the length of the text and the level of demand. Turnitin.com was designed to provide users with a simple process for submitting papers for a plagiarism test. A user completes a short form to submit the paper. The user identifies institution, department, course, name, and ID number. Then the user may submit the paper for testing by pasting a text-only document onto the Web page.

EVE2 (Essay Verification Engine), anti-plagiarism software that instructors may license for a one time fee and download to their hard drives. A user may submit a paper in .txt form, and the software will search the Internet, including term paper mills, for results. Although the software is targeted at all educational levels, most users seem to be high school teachers. As noted, an instructor must convert papers to .txt format and submit the text to the software. EVE2 then examines the papers and makes a large number of searches of the Internet to locate "suspect" sites. Once suspect sites have been located, EVE2 visits all of these sites to determine if they contain work that matches the paper in question. Searches require fifteen minutes to two hours for processing. This time is in part linked to the power of the user's hard drive. Once the search is complete, the instructor is given a full report on each paper that contained suspected plagiarism, including the percentage of the paper plagiarized, an annotated copy of the paper showing all plagiarism highlighted in red, and links to the plagiarized sites. The report does not distinguish which sentences have been plagiarized from which sites.

Knowledge Ventures is developing a suite of tools and middleware components for use in text to- text matching applications. CiteMaster, the first application to incorporate these tools, will use super–computer processing power to compare submitted text documents against the content of a proprietary academic database of textbooks, journals, Web content, and student-submitted papers. Through the use of customized algorithms and a parallel-processing platform, CiteMaster should be able to detect both verbatim and inexact text matches and to return results in real time. For each student-submitted paper, a dynamically generated "report" will be returned to a Web browser, including an HTML reconstruction of the original query document and relevant statistical information about the matches found. Matching or "suspect" sentences will be clearly identified in the text, with their proper citations listed accordingly.

CiteMaster is in the early stages of development, however, and is not yet commercially available.

PENALTIES FOR PLAGIARISM

1. Ask to resubmit the work and deducting total or a few marks for an assignment or course.
2. An action taken by the University/Institute authorities in the form of official warning that the next offence will be punished more hardly.
3. Withholding the scholarships/funds to the researcher.
4. A suitable disciplinary action which can be recommendation to cancel a degree, stopping the promotion, increment or suspension from the Institute/University.

AVOIDING PLAGIARISM

There is major problem with the areas and countries where English or the language in which the text is written is not a mother tongue or native language. The writer mostly tries to copy the matter when finds difficulty in language or ideas. Following points are mentioned by which the plagiarism can be avoided:

Paraphrasing

Books and articles should be used as a source of information from which one writes, in its own words. It is important to mention here that the source of information should always be acknowledged. In any written text one should rely on source texts for information, paraphrasing should be used the most. Before writing any text the writer should read the relevant material from different sources, understand the topic and then start writing in its own form and language. One should also explain the points, compare and contrast the views of different authors and add own comments on the topic under discussion. By doing these things one go beyond merely repeating the information which is found written already and makes a good assignment. The original ideas, work submitted should always be rewarded by the reader, checker/teacher.

Copying Directly from a Text

Many times one needs to copy or quote directly from a source material. In this case, certain rules must be followed:

1. Direct copying of part of a passage (*e.g.*, a whole paragraph) must occur rarely in a writing and once copied should not represent a large proportion of your own text. It should be clearly distinguished from the rest of your text in a way which makes it clear that it is a quotation.
2. Very short parts of a source text (*e.g.*, part of a sentence) can be copied when needed. However, the words that are copied must be immediately obvious to the reader. Care must be taken not to change any of the words.
3. The exact source of your quotation must be acknowledged. This must be done in a way which shows clearly how much is copied.

HOW TO PREVENT PLAGIARISM WHILE WORKING ON ASSIGNMENTS

1. Always put quotes in quotation marks and indent quotes of significant length so they stand out from the rest of the text. This will tell that this is taken from another source
2. Always acknowledge the source within the text and in full in the reference section.
3. In case of paraphrasing, an individual should always acknowledge the source of the ideas.

4. Make sure that the work is free from plagiarism before submitting.
5. The tutor/teacher must be asked for clarification if any doubts remain about the assignment.

9

Funding Agencies and Preparing Applications for Funding

FINANCIAL GRANTS

Research and development activities are considered as an integral component of higher education. In the present era of globalization and competition, every professional need financial grants from Government or private agencies/organisation to meet out the research needs and for the development of career through various other activities. There are many Government agencies which have different programs to provide financial grants. The persons receiving the financial assistance from the government are summarized in the flow chart given below in Fig. 9.1.

Basically, the financial grants are provided to the following:

1. **Individuals:** Professionals working in an Institute or University department or are retired. Students are also eligible to get scholarships and grants from different funding agencies through many schemes. For example Research promotion scheme (RPS) grant of AICTE.

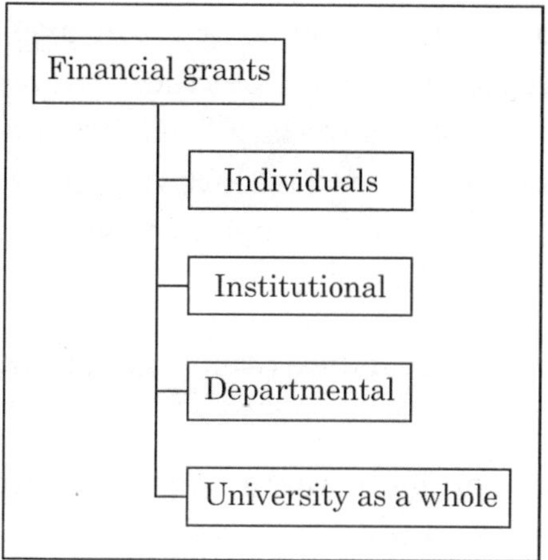

Fig. 9.1: Persons who can receive financial assistance from the government

2. **Institutional:** Institutes and colleges affiliated to some university are eligible to get grants. These grants are meant for the whole institute and not for any individual. For example: Industry- Institute partnership cell (IIPC) of AICTE.

3. **Departmental:** Some grants are meant for university departments. A team of few researchers work on a project sanctioned to the department or grant is sanctioned to develop some infrastructure in the department. For example UGC special assistance program (UGC-SAP) is given to a department.

4. **University as a Whole:** Few grants are sanctioned to the university as a whole to develop infrastructure or to develop research facilities. It becomes provocative of the recipient university how to distribute the grant to the needy departments. For example, UGC grant to enhance research facilities.

In the wake of the recent developments and the new demands that are being placed, it is necessary for a researcher and an institution to embark on some major research activities which have relevance to national needs and which will also be relevant for tomorrow's technology. But to meet out the expenditure for research activities, it is necessary to generate funds from a suitable source because most of the times a sound research project is not materialised due to shortage of funds. Major objectives which are met though financial grants are mentioned below:

- To promote research in newly emerging frontier areas/thrust areas.
- To selectively promote the general research capability in relevant areas taking into account capability of the host institutions.
- To encourage young scientists to take up challenging R&D activities.
- To give special encouragement to projects from relatively small and less endowed persons working in University Departments and Institutions.
- To encourage patenting facilities to scientists and technologists in the country for Indian and foreign patents on a sustained basis.

The above mentioned objectives set by the Government agencies are fulfilled through differently designed programs and schemes which are broadly mentioned here:

1. Travel grants are sanctioned to provide financial assistance to attend/present paper, deliver an invited talk and/or to chair a scientific session mostly out of country.
2. Research projects of individual researchers are funded by different agencies to meet non-recurring needs like equipments, infrastructure and recurring requirements like salary, contingencies, travel and consumables.

3. Different scholarships are given to upgrade the qualification of students to peruse higher degrees.
4. Fellowship programs are meant to encourage established scientists or eminent retired scientists.
5. Infrastructure funds are provided to upgrade infrastructure facilities of an institute/department.
6. Seminar grants are provided to organise a seminar/conference.
7. Funds are provided to organise Staff development programs, faculty development programs and refresher courses for the faculty.
8. Financial grants are also provided to enhance instrumental facilities.
9. Funds and grants are provided to projects in collaboration with industry or other institutes.
10. Many schemes of fellowships, grants, scholarships are in place to uplift special class of people.

There are many funding agencies that provide financial grants through different schemes. Some major Government agencies and their popular programs are mentioned here in Fig. 9.2.

DEPARTMENT OF SCIENCE AND TECHNOLOGY (DST)

Department of Science & Technology plays a pivotal role in promotion of science and technology in the country. The department has wide ranging activities ranging from promoting high end basic research and development of cutting edge technologies on one hand to service the technological requirements of the common man through development of appropriate skills and technologies on the other.

The Science and Engineering Research Council (SERC) of DST now the Science and Engineering Research Board (SERB) has the

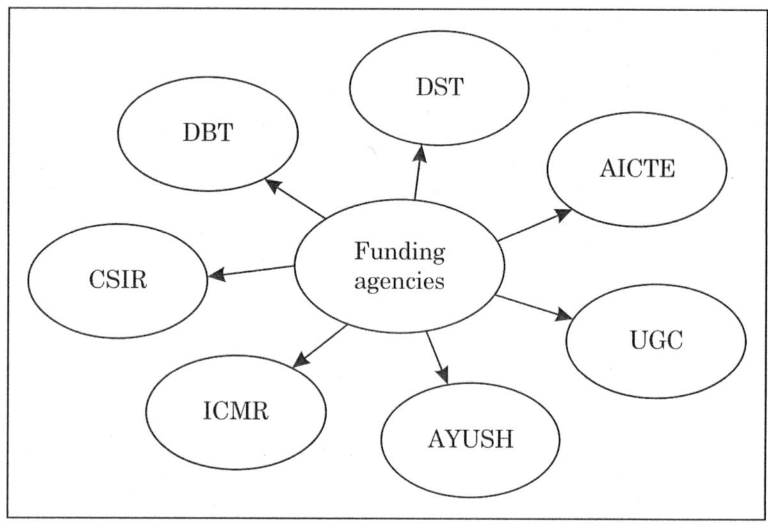

Fig. 9.2: Funding agencies that provide financial grants

broad mission of promoting and challenging new areas of science and engineering with focus on basic research in all disciplines. It also encourages involvement of academic and research institutions to undertake research in identified areas. It also acts as a mechanism for inter agency coordination with other S & T agencies for better management of R & D funding through joint/ complimentary funding in major projects.

Fund for Improvement of S&T Infrastructure in Higher Educational Institutions (FIST)

This is based on selective strengthening of the infrastructure for post-graduate education and research in emerging areas. Apart from support for basic equipment necessary for imparting quality teaching and modernization of laboratories provided to a large number of departments, support for creation of a few major facilities has also been provided.

Support is given for a period of 5 years. It is aimed at two levels:

- ***Level I*:** Moderate funding for improving quality of teaching and research through modernization of labs and for augmenting library facilities (from relatively small but active departments);
- ***Level II*:** Substantial funding for acquiring state-of-the-art equipments and setting up labs for conducting internationally competitive research (from well-established departments).

Sophisticated Analytical Instrument Facilities (SAIFs)

The Department of Science and Technology (DST) is providing facilities of sophisticated analytical instruments to researchers through its Sophisticated Analytical Instrument Facilities (SAIF) programme so that the non-availability of these instruments in their institutes may not come in the way of scientists in pursuing R&D activities requiring such facilities and they are able to keep pace with developments taking place globally. Thirteen Sophisticated Analytical Instrument Facilities (SAIFs) which provide sophisticated analytical instruments to users are functioning at approved centres. About 8,000 scientists/users are utilizing the facilities at the SAIFs annually. The SAIFs regularly organize short term courses/training programmes on use and application of various instruments and analytical techniques to create awareness among the users about them and on maintenance/repair/operation of instruments for technicians.

Human Resource Development and Nurturing Young Talent

Swarnajayanti fellowships scheme

Under this scheme a selected number of young scientists, with proven track record, are provided special assistance and support to enable them to pursue basic research in frontier areas of science and technology. The fellowships are scientist specific and not institution specific, very selective and have close academic monitoring. The support covers all the requirements for performing the research and includes a fellowship of Rs. 25,000/- per month for five years. Scientists selected for the award will be allowed to pursue unfettered research with a freedom and flexibility in terms of expenditure as approved in the research plan.

Fast track scheme for young scientists (FAST)

FAST track Young Scientists programme is aimed at providing quick research support to young scientists to pursue their bright ideas in newly emerging and front line areas of research in science and engineering. Under this scheme young scientists can submit proposals anytime of the year, subject to eligibility. Under this scheme, the upper limit for duration of the project is 3 years with total cost limited to Rs.23.00 lakhs (excluding Overheads). The Young Scientist not drawing any fellowship/salary are eligible for a lump sum fellowship of Rs. 35,000 per month.

Better opportunities for young scientists in chosen areas of science and technology (BOYSCAST)

The scheme provides opportunities to Young Indian Scientists and Technologies (up to the age of 35 years), to interact with international scientist community and institutions and to participate in research and training activities in frontline areas of science and technology. The duration of the Fellowship is 3-12 months. Annually, on an average, about 40 young scientists are awarded this fellowship in various areas. The selected young scientists are also encouraged to attend scientific conferences and visit other institutions in the host country. Candidates selected under this scheme are expected to further generate and spread their expertise at the S & T laboratories/ Institutes across the nation.

DST's Scholarship Scheme for Women Scientists and Technologists

The "Women Scientists Scheme (WOS)" has been aimed for providing opportunities to women scientists and technologists between the age group of 30-50 years who desire to return to mainstream science and work as bench-level scientists. Under this scheme, women scientists are being encouraged to pursue research in frontier areas of science and engineering, on problems of societal relevance and to take up S & T-based internship followed by self-employment.

The scheme is meant to encourage women candidates, preferably those having a break in career and not having regular employment. The selected candidates are given up to Rs. 35,000/- pm as Research Scholarship. The upper limit on Cost of these projects including the scholarship amount is Rs. 20 lakhs for a period of three years.

Assistance to Professional Bodies & Seminars/Symposia

The Department extends partial support on a selective basis, for organizing seminar / symposia/ training programmes / workshops / conferences at national as well as international level. The support is provided to Research Institutes/ Universities/Medical and Engineering Colleges and other Academic Institutes/ Professional Bodies who organize such events for the scientific community to keep them abreast of the latest developments in their specific areas.

Utilisation of the Scientific Expertise of Retired Scientists (USERS)

The main objective of the scheme is to utilize expertise and potential of large number of eminent scientists in the country who remain active and deeply motivated to participate in S&T development activities even after their retirement. The scheme has continued to play a significant role in involving a large number of retired scientists in S&T developmental activities. The main activity under the programme is preparation of books/monographs/state-of-the-art reports. The project duration is normally for two years and the Scientists awarded in this scheme are given an honorarium.

TECHNOLOGY DEVELOPMENT BOARDS (TDB)

The Technology Development Board is the first organization of its kind within the Government framework with the sole objective of translating the fruits of indigenous research into commercial products or services. The Board plays a pro-active role by encouraging commercial enterprises to take up technology oriented projects. Technology Development Board aims at accelerating the development and commercialization of indigenous technology or adapting imported technology to wider domestic application. The board provides financial assistance in the form of Equity, Soft loans, or Grants.

Fellowships

Ramanna fellowships

The objective of the Ramanna Fellowship is to offer continued core financial support to those researchers who have performed excellently well in their ongoing basic research projects of SERC.

Ramanujan fellowships

This fellowship is meant for brilliant scientists and engineers from all over the world to take up scientific research positioning in India, especially those scientists who want to return to India from abroad. The duration of the fellowship is valid for five years with a value of Rs. 50,000 per month for the first 3 years and Rs. 60,000 per month during the last two years.

JC Bose national fellowships

This fellowship is meant to recognize active scientists for their outstanding performance and contributions. The Department of Science & Technology administer this scheme. The fellowships are scientist–specific and very selective. The duration of the fellowship is valid for five years with a fellowship of Rs. 20,000 per month in addition to regular income.

ALL INDIA COUNCIL FOR TECHNICAL EDUCATION (AICTE)

The All India Council for Technical Education (AICTE) is a Statutory Body established by the Government of India through Act No. 52 of 1987 with a view to proper planning and coordinated development of Technical Education (TE) system throughout the Country. Several schemes have been initiated by AICTE for the development of research in technology including natural products research.

Research Promotion Scheme (RPS)

The objective of this scheme is to create and update the general research capabilities of the faculty members of the various Technical Institutes. The proposal should include a specific project theme with a clear statement of the objectives, details of equipments and other research facilities proposed to be acquired and the expected deliverables from the project. Funding to such projects is limited to Rs. 20 lakhs.

Career Award for Young Teachers

The scheme is open to regular teachers (who are not ad-hoc appointees or temporary) working in AICTE approved Technical Institutions. Teachers having post graduate degree as minimum qualification with consistently good academic career and a demonstrated aptitude with commitment to research work shall be eligible for the award. The age limit for the award is 35 years with relaxation of 5 years for women teachers. The award is for a period of three years.

Modernization and Removal of Obsolescence (MODROBS)

The scheme aims to modernize and remove obsolescence in the Laboratories/Workshops/Computing facilities (Libraries are excluded), so as to enhance the functional efficiency of Technical Institutions for Teaching, Training and Research purposes. The equipments are financed under the scheme up to a limit of Rs 20 lakhs.

Travel Grant

The Scheme extends financial assistance to the meritorious faculties of AICTE approved Departments / Institutions to present research paper in an International Conference/ seminar Symposium as per the AICTE Act. 100% grant is given in terms of air ticket, registration and per diem charges for the candidates below 35 years with a relaxation of 5 years for women candidates.

Staff Development Programme

The scheme intends to provide financial assistance to facilitate upgradation of knowledge, skill and intends to provide opportunities for induction training to teachers employed in AICTE approved Institutions.

Seminar Grant

The scheme provides financial assistance to institutions for organizing Symposium / Conference / Seminar / Workshop at National and International level in various fields of Technical Education. Upto Rs. 7 lakhs can be sanctioned depending upon the proposal. The scheme is also extended for registered professional societies.

Emeritus Fellowship

This scheme intends to utilize the services of highly qualified and experienced superannuated Professors at AICTE approved Institutions. The Fellowship is tenable for a period of two years extendable for one more year or up to the age of 70 years. The Fellowship consists of (i) honorarium for a duration of 2 years plus contingency charges.

Visiting Professorship

The main objective of the scheme is to supplement and provide expertise to teaching and research in those areas in which the host institute needs expertise. The Visiting Professorship is usually instituted in the emerging fields of Technical Education, where the host institute needs expertise. The maximum tenure of appointment of a Visiting Professor from within the country shall be one year and the minimum not less than three months.

Nationally Coordinated Projects (NCP)

To plan, coordinate and execute integrated Research and Development programmes by the established Institutions along with co-ordination of 3-4 Institutions.

National Facilities in Engineering & Technology with Industrial Collaboration (NAFETIC)

To establish sophisticated, testing, instrumentation and design facilities to industry in specialized/emerging areas of engineering and technology particularly, near industrial clusters.

Fellowships

National Doctoral Fellowship (NDF)

GATE Scholarship and PG Grant

Quality Improvement Programme (QIP)

UNIVERSITY GRANTS COMMISSION (UGC)

The University Grants Commission has been making proactive efforts to upgrade the knowledge and skills of faculty members in the institutions of higher education. The UGC has the unique distinction of being the only grant-giving agency in the country which has been vested with two responsibilities: that of providing funds and that of coordination, determination and maintenance of standards in institutions of higher education. Some of the important schemes meant for scientific technology are discussed here under:

Major Research Projects (MRP)

Research proposals based on thrust areas are invited under this scheme. The proposals are placed before the Expert Committee for screening. Principal Investigators of Major Research Project Proposals are invited for presentation before the Expert Committee. Sanctioning of amount depends upon the project.

Refresher Courses

The orientation programmes are intended to inculcate in the young lecturers the quality of self- reliance through awareness of the social, intellectual and moral environment as well as to discover self-potential and confidence. The orientation programmes contributes to the teacher awareness of the problems of the Indian society and the role of education, higher education leaders and educators in the resolution of these problems to achieve desired goals in national development.

Assistance for Strengthening of Infrastructure for Science and Technology (ASSIST)

The basic objective of ASIST is to assist selected science, engineering and technology departments in the universities which have already exhibited and achieved high quality performance to enable them to acquire such costly major equipment which cannot be approved out of normal university development grants, so that the attainment of excellence in post graduate education and research in the department is not handicapped due to the non-availability of such equipment. A total amount upto Rs. 100 lakh can be granted depending upon the project.

Research Awards

The Research Award scheme permits permanent teachers of universities and institutions full-time three-year research tenures in their respective field of specialization. Teachers who have a doctorate degree and have shown excellence in their fields are considered for the award. The Research Award is given to only those lecturers, senior lecturers, and selection-grade lecturers and readers who are in continuous regular service in a recognized institution on permanent posts and are under 45 years of age at the time of submission of their application. The awardee is eligible to avail the Research Award only once.

Research Workshops/Seminars/Symposia and Conferences in Colleges

This scheme provides financial assistance to institutions for organizing Symposium/ Conference/ Seminar/ Workshop at National and International level in various fields of Technical Education. Upto Rs. 5 lakh can be sanctioned depending upon the proposal.

Emeritus Fellowship

The University Grants Commission started the Scheme of Emeritus Fellowship in order to provide an opportunity to the superannuated teachers who have been actively engaged in research and teaching programmes. This scheme is valid upto 2 years for persons below 70 years. Fixed honorarium along with contingency is given to the selected persons.

Visiting Associateship

The Visiting Associateship scheme was introduced to enable the teachers of universities and colleges to visit institutions of advanced study and research for short periods with the aim of keeping them up to date with the developments in their areas of interest. It allows teachers of universities and colleges to visit institutions of advanced study with a view to keeping them abreast of the latest developments in their streams.

Faculty Improvement Programme (FIP)

The objective of the "Teacher Fellowship" under Faculty Improvement Programme is to provide an opportunity to the teachers of the Universities and Colleges to pursue their academic/research activities leading to the award of Master and Doctorate degrees. Contingency, Salary for substitute and travel grant is given under this scheme.

Part Time Research Associateship for Women

This scheme was initiated with the intention to provide opportunities to unemployed women with Ph.D. degrees, and with an aptitude for research, but unable to pursue the research work on regular basis due to personal or domestic circumstances. A fellowship along with contingency is given.

Student Fellowships

1. Junior Research Fellowship (JRF)
2. Senior Research Fellowship (SRF)

The objective of this scheme is to provide an opportunity to research scholars to undertake advanced study and research in engineering and technology, pharmacy subjects leading to Ph.D.

DEPARTMENT OF AYURVEDA, YOGA AND NATUROPATHY, UNANI, SIDDHA AND HOMOEOPATHY (AYUSH)

AYUSH is the acronym for Ayurveda, Yoga and Naturopathy, Unani, Siddha and Homoeopathy and includes therapies documented and used in these Systems for the prevention and cure of various disorders and diseases. India has a large infrastructure for teaching and clinical care under these Systems. The scientific validation of these therapies, however, still remains to be done on a wider scale. Based on above, AYUSH has initiated following programmes:

Scheme for Extra Mural Research (EMR)

The Department of AYUSH has introduced a Scheme for Extra-Mural Research in addition to the intra-mural research undertaken by the Research Councils for Ayurveda and Siddha, Unani, Homoeopathy and Yoga and Naturopathy. The Department has taken up a series of programs/interventions wherein evidence based support for the efficacy claims is needed.

The Department of AYUSH provides financial support for staff, equipment and contingencies (recurring and non-recurring) for the project over a period of 1-3 years upto a maximum of Rs. 30.00 lakhs.

Assistance for Exchange Programme / Seminar / Conference / Workshop on AYUSH

The scheme is aimed at achieving the objective of Promotion and development of *AYUSH* with special reference to their strengths and contemporary issues and development and dissemination of proven result of R&D work in *AYUSH*; and to provide a forum where horizontal and vertical interaction among stake holders of *AYUSH* can take place through International Cooperation, Conference and Seminars (International, National & Regional).

Under the scheme funds are provided to eligible organizations for meeting expenses on air fare, boarding and lodging, local transport and other contingencies of the delegates invited from abroad to attend the international Conference on *AYUSH*.

1.	National conference/Workshops / Seminar organized by Department of AYUSH	Rs. 3.00 lakh to Rs. 5.00 lakhs as per requirements.
2.	National Conference / Workshop / Seminar organized by the State Government	Maximum Rs. 3.00 lakhs
3.	National Seminar organized by NGOs	Maximum Rs. 1.00 lakh.
4.	National Seminars or Workshops / Conference by eminent Institutions / University	Maximum Rs. 2.00 lakhs.

Scheme for Grant-in-aid to Non-profit/non-governmental AYUSH Organizations Institutions for Upgradation to Centre of Excellence

To support creative and innovative proposals for the up gradation of both functions and facilities of reputed AYUSH institutions to levels of excellence. Upgrading functions implies adding new long-term functions or making significant qualitative improvements in the existing functions. It also includes support for human resources who will be attached to the new functions.

DEPARTMENT OF BIOTECHNOLOGY (DBT)

Department of Biotechnology (DBT) is an Indian Government department under the Ministry of Science and Technology responsible for administrating development and commercialization in the field of biology and biotechnology in India. The department has initiated various programs and schemes for the development of the scientific activities which are detailed below in Fig. 9.3 and 9.4 respectively.

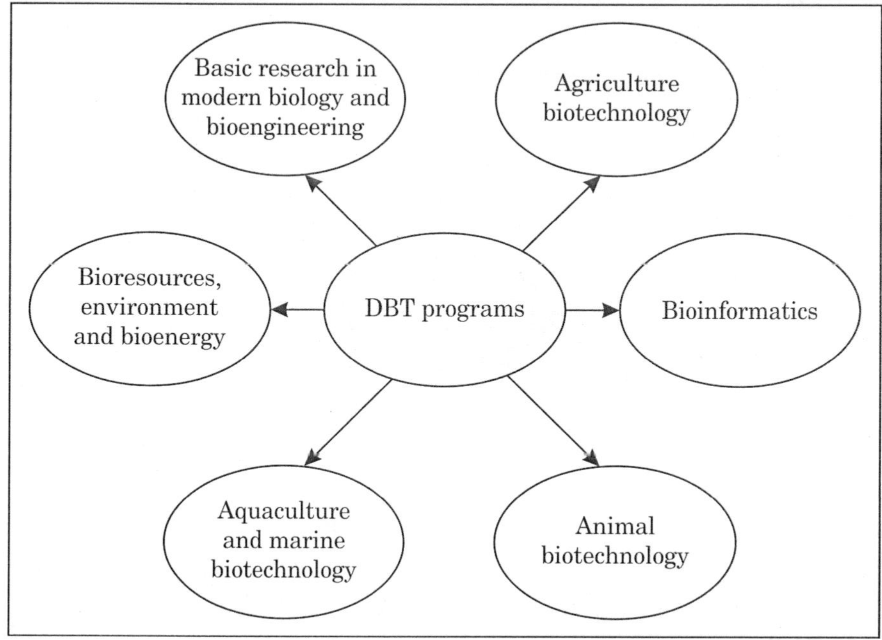

Fig. 9.3: DBT programs for the development of the scientific activities

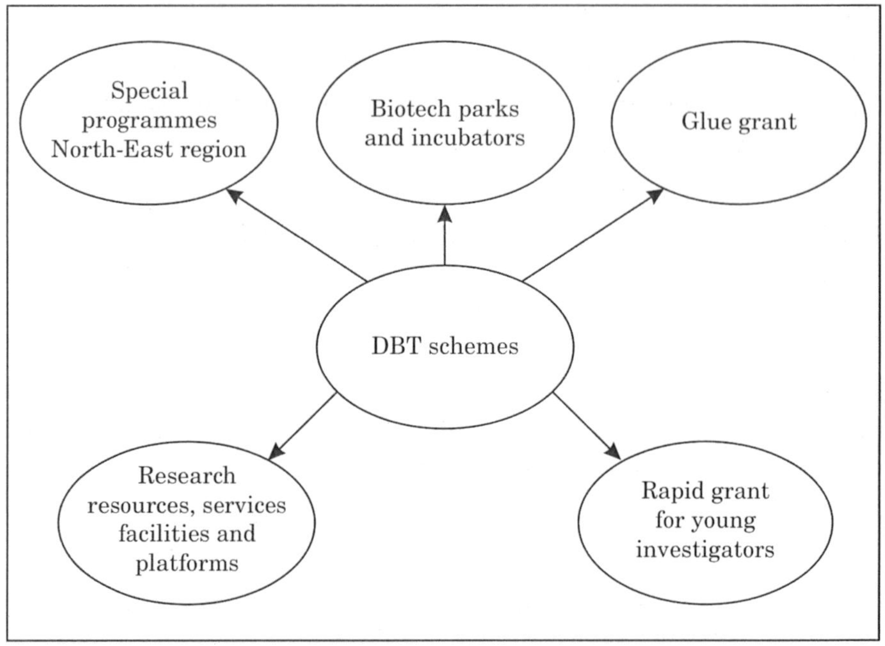

Fig. 9.4: DBT schemes for the development of the scientific activities

Fellowship Programs of DBT

DBT biology scholarships for 10+2

DBT biology scholarship is granted to 100 students from schools around the country. The students must be of higher secondary, intermediate or 10+2 level from any of the boards—CBSE, ICSE or state boards. This is done to encourage students to pursue studies in biological sciences and related subjects after 10+2. A total of 44 students from CBSE board, four from ICSE board and 52 (two from each state board) from the various state boards are selected for the award which carries a cash prize of Rs 20,000, a certificate and a medal. The selection is on the basis of merit list provided by the boards.

DBT - JRF

The department provides fellowships to biotechnology students to pursue doctoral research in universities/research institutions across the country. The students are selected through Biotechnology Eligibility Test (BET) now being co-ordinated by the National Centre for Cell Sciences.

DBT-Research Associateship (DBT - RA)

The department provides fellowships for post doctoral research in frontier areas of biotechnology and life sciences at premier institutions in India. The DBT-RA programme is being implemented by the Indian Institute of Science, Bangalore. The fellowship is initially awarded for a period of two years and the support can be extended for 1 to 2 years based on review of progress.

Khorana program for scholars

This program of scholarship is an Indo-US collaborative effort to create an effective scientific contact between the biotechnology students of India and the biotech students of the United States of America. The scholarship allows Indian students of B Tech, M Tech and M Sc to undertake research internship in Wisconsin-Madison University of USA. The programme has now been expanded to some other universities of USA like Minnesota, Michigan, Iowa, Indiana and Georgetown.

Award Schemes run by DBT

National bioscience award for career development

National Bioscience Award for Career Development (NBACD) recognizes outstanding contributions of young bio-scientists below 45 years through a grant for research projects to help in their career development. The award aims to boost outstanding research in basic and applied biosciences can be granted to 10 candidates for a given financial year. It carries a cash prize of Rs 1 Lakh, a citation and Rs 3 Lakhs/year as research grant for three years.

National women bio-scientist awards

The Department of Biotechnology (DBT) recognizes the contributions of senior and young women scientists in the country who are working in the areas of biology and biotechnology and have made outstanding contributions in research, technology and product development through two separate awards. Nominations are invited for the awards through advertisements.

Biotech product and process development and commercialization awards

Upto five awards are given every year for biotech products and process development and commercialization. Each award will carry a cash amount of Rs 2.00 lakhs along with a citation. Rs 5.00 lakhs would be given if the product is commercialized and has much higher impact of utilisation in the country. Nominations are invited for the awards through advertisements.

The innovative young biotechnologist award (IYBA)

The Innovative Young Biotechnologist Award (IYBA), initiated in 2005, is a career-oriented prize to identify and nurture outstanding young scientists with innovative ideas and desire of pursuing research in biotechnology. The prize is for those below 35 years of age. This includes scientists without regular employment. Nominations are invited for the awards through advertisements

INDIAN COUNCIL FOR MEDICAL RESEARCH (ICMR)

The Indian Council of Medical Research (ICMR), New Delhi, the apex body in India for the formulation, coordination and promotion of biomedical research, is one of the oldest medical research bodies in the world. Various schemes run by ICMR are as follows:

- Intramural Research
- Extramural Research
- Travel grant
- Seminar/symposia/workshop grant
- Research fellowship
- Post doctoral research scheme
- Centre for advanced research
- Task force project

COUNCIL FOR SCIENTIFIC AND INDUSTRIAL RESEARCH (CSIR)

Council of Scientific and Industrial Research (CSIR) established in 1942, is an autonomous body and India's largest research and development (R&D) organisation, with 37 laboratories and 39 field stations or extension centers spread across the nation, with a collective staff of over 17,000. Although CSIR is mainly funded by the Ministry of Science and Technology, it operates as an autonomous body registered under the Registration of Societies Act of 1860.

CSIR contributes immensely to the fostering, sustenance and upgrading of the human S&T resources of the country and its support umbrella covers all age groups ranging from sixteen to sixty-five, in addition to adopting human resources management systems that improve its ability to build successful, high performance organizations. It is CSIR's contribution to generation of the pool of bench-level workers in science that distinguishes it from other funding agencies.

Shanti Swarup Bhatnagar Prize for Science and Technology

The Shanti Swarup Bhatnagar (SSB) Prizes are awarded annually by the Council of Scientific and Industrial Research (CSIR) for notable and outstanding research, applied or fundamental, in biological, chemical, earth, atmosphere, ocean and planetary, engineering, mathematical, medical and physical sciences. The prize comprises a citation, a plaque, and a cash award of Rs. 500,000. In addition recipients also get Rs. 15,000 per month up to the age of 65 years.

Apart from this CSIR have programs for financial grants to the research projects and other fellowships programs for the research scholars.

India is one of the top-ranking countries in the field of scientific research. Indian Science has come to be regarded as one of the most powerful instruments of growth and development, especially in the emerging scenario and competitive economy. The present need is to utilize most of the Government schemes in an effective manner to encourage research in thrust areas so that the research findings may lead to the effective implementation for the upliftment of the society.

PREPARING AN APPLICATION FOR FUNDING

Funding agencies receive huge amount of applications from researchers/professionals. These applications are generally screened by concerned experts and are rejected or recommended for next or final round. If not specifically and professionally prepared, the applications are rejected easily and thus the chances of getting financial grant for that particular project become nil. It is therefore necessary to prepare the application well to get financial grant from any funding agency.

General Tips

There are some general tips which should be kept in mind while preparing a proposal.

1. Every funding agency has its own format for each scheme. Use the appropriate format.
2. The selection of funding agency and a particular scheme should be done carefully according to needs.
3. Every funding agency has its own priority areas; therefore, the priority areas should be carefully examined and applied accordingly.
4. For few schemes the applications can be made throughout the year while for others the applications are invited. So the application must be made accordingly.
5. Your application must reach well before stipulated time along with the necessary enclosures and fee, if any.
6. For travel grant and seminar grant, the application must be submitted at least 3 - 4 months in advance.
7. Most of the funding agencies now invite online applications. So the application must be submitted through their online proposal submission portal.

PROPOSAL DEVELOPMENT

The proposal should be developed strictly as per the required format. However, a typical proposal contains few basic details which are required to be filled up during the application.

Title

The title of the proposal is the first thing which appeals an expert who scrutinizes/ screen or evaluate the proposal. The title should not be very lengthy and possibly be kept concise but it should reflect the main objective of the proposal. It should not contain any abbreviations or any special character. For example: Novel microwave extraction of *Tinospora cordifolia* using central composite design.

Objective

The main objective of the project must be clearly mentioned in this section. The objective or objectives must be designed in such a way that it should reflect main aims of the project.

Rationale

In this section, the purpose of the project should be clearly defined. Proper justification if needed, that why the project is selected and how it can provide additional information to the existing one.

Background

In this section, the introduction and background for the proposal is written in detail. The section contains proper bibliography. Care must be taken that only specific and relevant information be given in this section. This is the section which provides proper justification of preparing a project and reflects the need of the present project. In this section, the gaps in the previously conducted studies are identified and project the need to fulfill them in the current project.

Work Done so Far

In this section, literature survey or the studies conducted so far in the relevant areas of the selected project are mentioned. The written text in this section should be supported by suitable bibliography. Few funding agencies separately ask the studies conducted at national and international levels.

Expected Deliverables

In this section, it should be clearly mentioned that what are expected achievements of the project. The expected commercial and social outcomes must be defined in this section.

Methodology and Work Plan

In this section, the proper work plan of the project must be explained in detail as shown in Fig. 9.5. At least quarterly or half yearly targets should be mentioned. The methodology should contain brief details of the instrumental and experimental methods, so that the evaluator should be able to clearly understand the way the project is going to be carried out. A sample copy of the activity chart which is presented in a project proposal is presented here.

Activity	*Quarterly (3 monthly) details*												
	0	*1*	*2*	*3*	*4*	*5*	*6*	*7*	*8*	*9*	*10*	*11*	*12*
Receive Grant													
Purchase of equipment, plant material, glassware and chemicals													
ChEI studies with microplate method using ELISA technique													
Column chromatography of the potential extracts/ fractions													
ChEI studies with isolated compounds													

Fig. 9.5: (*Contd...*)

Fig. 9.5: *(Contd...)*

Activity	*Quarterly (3 monthly) details*												
	0	***1***	***2***	***3***	***4***	***5***	***6***	***7***	***8***	***9***	***10***	***11***	***12***
Structure elucidation by spectroscopic, spectrometric & X-ray diffraction methods, molecular docking studies													
Data interpretation, compilation, final report and communication of research papers													

Fig. 9.5: Sample representation of a work plan of the project

Budgeting

This is a very important section and must be prepared carefully. The upper limit of the funding agency should be kept in mind and as per the need of the project the budget details should be prepared. This section contains details of the recurring and non- recurring expenses. In non- recurring section, year wise details of the infrastructure development and instruments to be purchased should be mentioned while in recurring budget section, the year wise budget of the consumables, fellowship/salary, and travel, contingency and overhead expenses should be mentioned. Proper justification of the each head must be given. A sample of budgeting is presented herein Figs. 9.6. and 9.7

Sample

BUDGET ESTIMATES: SUMMARY

Item	*BUDGET*			*(in Rupees)*
	1st year	*2nd year*	*3rd year*	*Total*
A. Recurring				
1. Salaries/wages	3,16,800.00	3,16,800.00	3,16,800.00	9,50,400.00
2. Consumables	1,00,000.00	1,00,000.00	1,00,000.00	3,00,000.00
3. Travel	50,000.00	50,000.00	50,000.00	1,50,000.00
4. Other costs/ Contingency	1,00,000.00	50,000.00	50,000.00	2,00,000.00
B. Equipment (HPLC)	19,00,000.00	19,00,000.00		
Grand total (A+B)	35,00,400.00			

Justification: HPLC instrument will be used for the separation and analysis of phytoconstituents.

Fig. 9.6: Sample representation of the budget estimates

Sample

BUDGET FOR SALARIES / WAGES

BUDGET		*(in Rs.)*			
		1st yr (12 months)*	*2nd yr (12 months)*	*3rd yr (12 months)*	*Total (36 months)*
Designation & number of persons	Monthly emoluments				
Research Associate-1	26,400.00 (24000 + 10% HRA *i.e.*, 2400 as per university rule)	3,16,800.00	3,16,800.00	3,16,800.00	9,50,400.00
Total	26,400.00	3,16,800.00	3,16,800.00	3,16,800.00	9,50,400.00

Fig. 9.7: Sample representation of the budget estimates for salaries/wages

Investigators

In this section, the details of the principal investigator (PI) and co-investigators (Co-PI) are included. Proper bio-data in the required format should be furnished. The investigators should definitely mention the details of the previous publications / achievements in the relevant area of the project which will reflect that the investigators have sufficient expertise to carry out the project applied. Because most of the funding agencies prefer that the investigators should have worked in the relevant discipline earlier also.

10

IPR and Filing a Patent

INTELLECTUAL PROPERTY RIGHTS (IPR)

Intellectual property (IP) refers to creation of mind such as design, process, symbol and scientific inventions which have commercial values. The inventors of these things are protected by some rules and regulations known as Intellectual property rights (IPRs). The main objective of these regulations is to respect creativity and progress of innovation.

TYPES OF INTELLECTUAL PROPERTY

There are various types of IP like

Copyright	Patents
Trade marks	Industrial designs
Trade secrets	

Copyright

Copyright is legal rights protected by the creators or inventors for their literary, artistic work or technical drawings. For example, a copyright can be obtained for a flow chart drawing created to show a process by some scientist. The other person cannot copy without written permission of the inventor or creator.

Patent

This topic is discussed later in detail in this chapter.

Trademarks

A trademark is a sign or design to distinguish or to identify a particular type of products. For example Dettol is a trademark by which one easily identifies the products. The manner in which Dettol is written is a registered trademark.

Industrial Designs

An industrial design is an aesthetic aspect of a product. This may be two or three dimensional design of an article.

Trade Secrets

Trade secrets are some formula to prepare a particular type of recipes or something related. It can be in the form of a formula, process or practice.

Infringement

Unauthorised use of intellectual property rights is called Infringement. The infringement can be of copyright by copying the protected matter, patent by disclosing the patented object details, trademarks by making similar or resembling objects and designs and can be of trade secrets by disclosing the formula etc. Infringement is liable to be prosecuted under law.

PATENT

A patent is legal rights granted to a person who lead to an invention and is useful for the existing information. The exclusive rights given to the inventor or the person to whom patent is granted *i.e.*, patentee is given rights for a limited period. These legal rights prevent others to use, sell or making that invention for that period of time. But after the expiry of duration of patent *i.e.*, 20 years, anybody can make use of patent/invention. The invention then becomes part of public domain. The main requirement for an invention to be patented is that the invention must be a new piece of work or a new product or process involving inventive steps and should be of any industrial application. Inventive step is defined as feature of invention which includes some new technical procedures that have never taken place or existed and have potential of some economic significance. Registration is a prerequisite for patent protection and the protection granted is territorial in nature *i.e.*, patent granted in a country will give the owner of the patent right only within that country. Patent act 1970 as amended in years 1998 and 1999 along with patent rules 1972 governs patent in India.

HOW TO OBTAIN PATENT IN INDIA

First check whether the invention is patentable and the matter does not fall in list of non-patentable inventions. If it is patentable, the invention must meet three criteria of patentability *i.e.*, novelty, non-obviousness and industrial applications.

The invention is thoroughly checked whether the invention is really new by performing a "Prior Art Search". This step is necessary to confirm that the claims made in the patent are new and has not already been done by someone, somewhere in the world.

PATENT PROCESS IN INDIA

An application for patent must be made in format prescribed by the Patents act and Rules. All applications must be accompanied by detailed patent specifications clearly describing the invention claimed. In India patent application comprises of four forms *viz*, form 1, form 2, form 3, form 5 and form 18. These forms are not available commercially but the formats are standard and can be obtained from the website. While forms 1, 3, 5 and 18 are simple consisting of just 1 or 2 pages, form 2 is the one which constitutes the "heart of patent application".

The Patent Application of an inventor has following main stages:

Drafting Filling

Publication Examination

Opposition Grant

Drafting

The process of drafting a patent application is the most important work in getting a patent because the whole process of granting a patent depends on it. There are standard formats in which the application is made and a few necessary points are required to be discussed in the application. Most of the researchers take help of patent agents who are well versed with the process and to assure that the application is not rejected. The patent agents obtain power of attorney from the inventor and then communicate to the patent office on behalf of the inventor.

Filling

An application for a patent can be filed by the true and first inventor. It can also be filed by the assignee or legal representative of the inventor. If the application is filed by the assignee, proof of assignment has to be submitted along with the application. The applicant can be national of any country.

Every application shall be accompanied by a provisional or complete specification. Provisional applications are generally filed at a stage where some experimentation is required to perfect the invention. Provisional application is made when some innovative idea strikes into the mind of a researcher and immediately the provisional application is made to protect the legal rights at that stage itself. Provisional application does not contain experimental details.

A provisional specification shall contain:

a. Title

b. Written description

c. Drawing, if necessary and

d. Sample or model if required

However, the complete specification shall contain experimental details and claims thereof as follows:

a. Title

b. Abstract

c. Written description

d. Drawings (where necessary)

e. Sample or model (if required by examiner)

f. Enablement and best mode

g. Claims and

h. Deposit (microorganisms)

a. Title

Generally, a word or a phrase indicating the content of invention is called the Title. The title of the invention should be written in such a way that it should create curiosity among the interested persons to go into the detail. The impressive title helps to get the commercial utilization of the patent.

b. Abstract

This is a short paragraph describing the invention in precise manner. Abstract is also important as the sorting of inventions is made on the basis of abstracts and a good written abstract has more chances to get wider attention.

c. Written description

It is an important part of specification. It contains the complete and elaborate description of the invention. Description generally starts with a background of the invention. The written description explains the invention clearly and comprehensively, with the help of examples, drawings and models, where and when required.

d. Drawings

Written description must be supplemented with drawings, where and when required because flow charts, drawings shows impressive details of the project. Care must be taken that the drawings should be clearly labelled.

e. Samples or models

On initiative of the inventor or when required by the patent examiner, samples or models might be submitted to patent office. Such samples or models will provide better understanding of the invention.

f. Enablement and best mode

The applicant should explain the invention in a simple and detailed way that it should enable a person with ordinary skill to understand the invention. It should not only enable but the applicant should also describe the best mode of carrying out the invention.

g. Claims

They are most important elements in specifications. Claims define the abilities and limits of the utilization of the invention. If an invention involves microorganisms, which can't be described by writing then a sample of microorganisms has to be deposited at an internationally recognized depository. There is an internationally recognized depository at Chandigarh.

A provisional specification can't be filed if an application has been filed in foreign country (Convention Country) before the Indian filing and if the application is PCT application (PCT application is detailed later in this chapter).

In case of provisional application, it is necessary to file complete specifications within twelve months (extendable to fifteen months) of filing the provisional specification. Each specification should contain only one invention. If there is more than one invention in a specification, separate applications have to be filed for each.

Cost of Filing Application

Presently, the official fee for filing an Indian Patent application is Rs 1000/- (for individuals) and Rs 4000/- (for corporate/organisms). Request for examination on Form 18 has to be filed for the application to be examined. Fee for examination is Rs 2500/- for individuals and Rs 10,000/- for corporate. Thus, as an individual filing expenses on account of fee alone are Rs 3500/- while that for corporate it will be

Rs 14,000/-. Attorney fee for drafting and filing generally ranges from 10,000-50,000/-.

Priority Date

It is the date of first filing allotted by the patent office to an application. If a provisional application is followed by complete application. The priority date shall be filing of the provisional application. If the Indian application is filed after a foreign or PCT application, the priority date shall be the date of filing of the foreign or PCT application. If the application is divided into two parts, the priority date shall be date of filing of the parent application.

Priority date is the date of reference used by the patent office to determine the newness of the invention. If the claimed invention is a part of public knowledge before the priority date, it will not be eligible for patent.

Place of Filing

Patent application can be filed at any of the four patent offices in India. Patent offices are located at Kolkata, Delhi, Chennai and Mumbai.

Documents to be Submitted at Time of Filing

The following documents have to be submitted at the time of filing of patent application:

a. Form 1-Application for grant

b. Form 2-Provisional or Complete Specification

c. Form 3-Statement and undertaking by the applicant

d. Form 5-Declaration as the inventor ship

Publication

A patent application will be published in the official Gazette of patent office of India on expiry of eighteen months after the priority date. It can be published earlier, if such a request is made by the applicant. The application will not be published if directions are given for secrecy, until the term of those directions expire. It will also not be published if the application is withdrawn three months before publication date. On publication specifications are made public.

Examination

After publication of the patent application, the next step is to examine the application contents. A request has to be made by submitting the prescribed format to the patent office in this regard.

Request for Examination

The process of Examination starts with a request for examination. The request has to be made within 36 months from the date of priority or filing.

Examination

On receiving the request, the controller shall direct the patent application to the Examiner for examination. To start with, the examiner makes a formal examination by verifying propriety and correctness of all documents filed with application. Later, verification of the patentability of the application is done. The patentability analysis includes all the patentability requirements. After confirming that the application falls within the scope of patentable subject matter, the examiner conduct a prior art search to check if there is prior art, which anticipates the invention claimed. Prior art search for anticipation includes search for anticipation by publication, filing of complete specification etc. He then verifies the existence of invention step, industrial application and Enablement and Best Mode. The examiner will then give the examination report within 1 month from date of reference by controller and that term shall not exceed three months. If the examination report is adverse the controller sends a notice to the applicant and gives him an opportunity to amend the application in order to proceed further. If the applicant does not make such changes, the application must be rejected. The controller has the power to divide the application, post date the application, substitute applicants and reject the application. An order of division will be given if the application contains more than one invention and if it is required to file separate application for each invention. The application might be post dated to a period of six months if requested.

Substitution of inventors is done if the invention is wrongfully mentioned or if a joint invention has not been mentioned in the application. The controlled has the power to reject the application if the applicant does not comply the reply of best objection.

Opposition

Pre-grant opposition

Any person can file an opposition for grant of patent after the application has been published.

Opposition may be filed on any of the following grounds:

a. Non-compliance of patentability requirements

b. Non-disclosure or wrongful disclosure of genetic resource or traditional knowledge.

Post-grant opposition

Any person can file an opposition within period of 12 months after the grant of patent. It can be filed based on following grounds:

a. Wrongful obtainment of invention by inventor

b. Publication of claimed invention before priority date

c. Sale or import of invention before priority date

d. Public use or display of invention

e. The invention does not satisfy patentability requirements

f. Disclosure of false information to patent office

g. Application for the invention is not filled within 12 months from the date of convention application

h. Non-disclosure or wrongful disclosure of the biological source

i. Invention is anticipated by traditional knowledge

Process of opposition

On receiving a notice of opposition, the controller notifies the patentee. He then constitutes an opposite board to deal with opposition. The opposition board decides the issues after giving reasonable opportunity of hearing to both parties. The opposition board might invalidate the patent, require amendments or maintain the status quo. If amendments are required, they are made within prescribed period in order to maintain the patent.

PATENT GRANT

If the application satisfies all the requirements of the patent act, the application is said to be in order for grant. An application in order for grant shall be granted expeditiously. A granted patent shall be published in official gazette and shall be open for public inspection. Every granted patent shall be given filling date. The patent will be valid throughout India. A granted patent gives the patent holder the exclusive right to make, use, sell, and offer for sale and import the product or use the process. However, the government can make use of the patent for its own purposes or for distributing an invention relating to medicine to hospitals and dispensaries. Furthermore, any person can make use of the patent for experiment or education. Formats required for filing a patent with details of specifications and request for examination are given as sample. Communication from patent office in the form of queries raised and final patent grant is also shown as sample in Appendices.

INTERNATIONAL PATENTING

One of the critical aspects of patents is that patents are valid only in the country they are filed and nowhere else. If a researcher feels that his invention is good, has lots of commercial value then he can go for international patenting in several countries. In order to secure rights in multiple countries one must file PCT application. PCT refers to Patent Corporation Treaty, initiated in year 1970, it is an agreement signed by more than a hundred countries, to facilitate the filing of international patents by inventors. It is administered by World Intellectual Property Organization (WIPO). PCT implements the concept of single international application, valid in member countries of PCT. The number of countries signatory to PCT now stands at nearly 140. Apart from simplification of the patent filing procedure, another aim of PCT is to promote easy exchange of technical information between different countries by providing a common platform for publication of patent applications from different countries. Details about PCT are available at www.wipo.int/pct/en/.

11

Technology Transfer, MoU and Confidential Agreements

TECHNOLOGY TRANSFER

"Technology transfer" is defined as a procedure which includes the transfer of any process, skill, method along with documentation and professional expertise between two parties, organisations and institutions to ensure scientific and technological developments to create new products, processes, applications or materials.

It is the process by which existing knowledged facilities or capabilities are utilized to fulfill public and private needs. It includes a range of formal and informal cooperation between laboratories and the public and private sectors.

Technology transfer process is utilized to commercially exploit the new discoveries and many times two organisations come together to work jointly when a single organisation does not have skill, expertise or resources to carry out the project alone.

Technology Transfer Activities

Technology transfer include

1. Processing and evaluating invention disclosures.
2. Filing of patents.
3. Technology marketing property arising from research activity.
4. Assisting in creating new businesses and promoting the success of existing firms.

The result of these activities will be new products, more high-quality jobs and an expanded economy.

Concept of Technology Transfer

The concept of technology transfer is to get the ideas, invention and technologies developed with tax amounts into the hands of the private sector as quickly as possible in a form useful to that community. The idea is to get the private sector involved in the development of a technology and products at an early stage so that the end result is a useful new product or service offered on world markets.

Purpose of Technology Transfer

The main purpose of technology transfer is to strengthen the economy by accelerating the application of laboratory technology and resources to private and public needs and opportunities.

Reason for Technology Transfer

1. To encourage use of technology developed using taxpayer dollars to benefit society.
2. To demonstrate research program relevancy and values.
3. To demonstrate Federal researchers to partner with the

private sector, leverage resources and share ideas in a protected environment.

4. To increase visibility to researchers and enable them to generate and earn royalty income.

Importance of Technology Transfer

1. To elucidate necessary information from R and D to actual manufacturing.
2. To elucidate necessary information of existing products between various manufacturing places.
3. To exemplify specific procedures and points of concern for the two types of technology transfer to contribute to smooth technology transfer.

Technology Transfer Process

Technology transfer is the process of transferring discoveries and innovations resulting from university research to the commercial sector and typically comprise of several steps as depicted in the flow chart given below in Fig. 11.1.

Technology Transfer Team

Many organisations now formulate a technology transfer team which consists of the following members:

1. R and D process technologist.
2. QA representative
3. Production representative
4. Engineering representative
5. QC representative

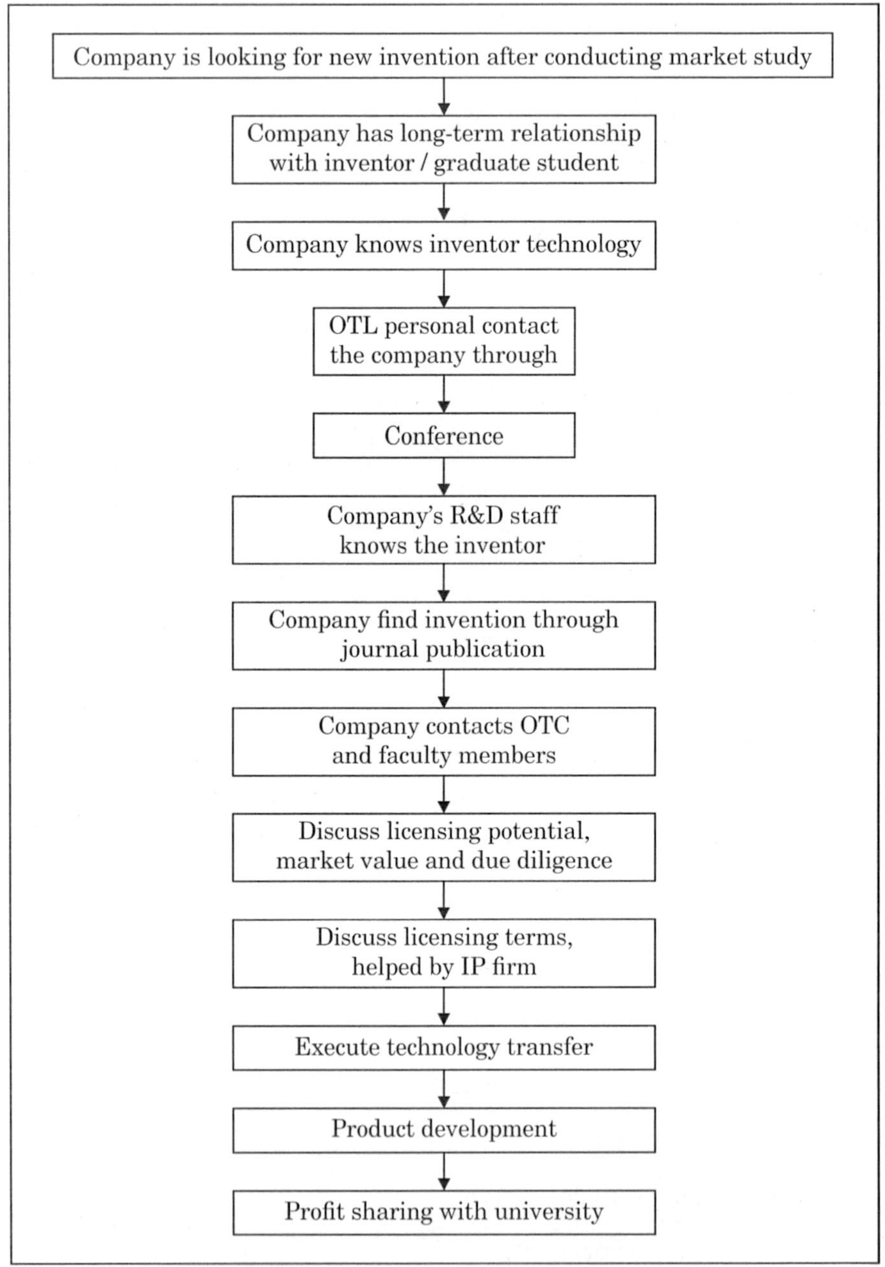

Fig. 11.1: Steps involved in technology transfer

Technology Transfer Documentation

The technology transfer documentation includes:

1. Organization for technology transfer
2. R and D report
3. Product specification file
4. Technology transfer report
5. Verification of results of technology transfer.

Technology Transfer Checklist

It consists of

1. Production master formula
2. Manufacturing instructions
3. Dispensing instructions
4. Analytical methods
5. Previous process validation
6. Previous analytical validation
7. Cleaning instructions and validation
8. Stability reports
9. Excipient specifications and source
10. Active specifications and source
11. Primary packaging material specifications and source
12. Packaging instructions
13. Process deviations file
14. Analytical deviations file

15. Reject and rework file
16. Specimen manufacturing batch record

MEMORANDUM OF UNDERSTANDING (MOUs)

Memorandum of understanding (MoU) is a document of agreement between two or more parties to decide a common line of action. MoU is put in place to establish a clear understanding of how the deal will practically function and each party's role and compensation.

According to Mr. Elliott, "in the memorandum of understanding, simply set out the details of the deal, what everyone agrees to do and how and when everyone will get paid, what percentage do you add taxes, etc.

When two or more organisations commonly decide to work on a certain project, a document is signed between them to decide the roles and responsibilities of each party and how the project will be carried out. The terms and conditions are mentioned so that one should have clear idea about its role and responsibilities.

List of Aspects Included in MoUs

1. The date of the MoUs
2. Describing the situation of the parties involved and how they relate to each other.
3. What services each party contributes to the deal before, during and after the joint venture.

Communication Details

1. The names and contact information of each other.
2. Any probationary or trial period (MoUs is signed initially for five years and may be renewed by mutual agreement between

the parties as in case of the United Nations Industrial Development Organization (UNIDO))

3. Any set dates to review activity, performance, or satisfaction with deal.
4. What parts of this deal are open to change and how. (this is done by written agreement between the parties)
5. What aspects of the deal should require formal notification and how.
6. How disputes will be settled. (in case of UNIDO the parties will use their best efforts to settle promptly such dispute through direct negotiation)

Compensation Details

1. Who handles the money and how
2. How people are paid (who pays who, by what method, in what currency, deposited where)
3. When people are paid (the same day, every month, immediately after the transaction)
4. How much people are paid (flat fee, a percentage of the sale if so, does this include GST, a percentage of the profit and if so, what are the applicable costs and how much are the on all customers, on certain customers and if so, how is sales are tracked and reported).
5. How long people are paid (for the initial sale of a customer, for the lifetime of the customer's business, for the duration of the contract, for 6 months after the contract ends).

Terms of Agreement

1. When the deal starts (on a certain date, during a limited even

as soon as a sale occurs). MoUs will enter into force upon signed by the parties.

2. How long it lasts (for a certain period, indefinite until someone ends, at the end of an event).
3. How the deal is terminated (by one or the both parties, under what circumstances, how is the end point is carried out). Each party shall have the right to terminate the MoU as per the terms and conditions decided in the agreement.
4. What happens at the end of or after the deal.

Miscellaneous

1. Any restrictions to either party.
2. Any disclaimer statements.
3. Any privacy statements (such as revealing the sales amount but not information about the customers)
4. A place for all parties to sign the agreement.

CONFIDENTIAL AGREEMENTS

A non-disclosure agreement (NDA), also known as a confidential agreement (CA), proprietary information agreement (PIA) or secrecy agreement, is a legal contract between at least two parties that outlines confidential material, knowledge or information that the parties wish to share with one another for certain purpose, but wish to restrict access to or by third parties.

Confidential agreement creates a confidential relationship between the parties to protect any type of confidential and proprietary information or trade secrets.

Confidential agreements are commonly signed when two companies, individuals or other entities are considering to do business and need to understand the processes used in each other's business for the purpose evaluating the potential relationship.

Types of Confidential Agreements

They are mainly of two types:

1. Many confidential agreements are unilateral or one-way agreements, where one party wants to disclose certain information to another party but needs the information to remain secrets for some reason, perhaps due to secrecy requirements required to satisfy patent laws or to make sure that the other party does not take and use the disclosed information without compensating the disclosure.
2. Another type is a mutual agreement, where both parties will be supplying information that is intended to remain secret. This type of agreement is common when businesses are considering some kind of joint venture.

Content

1. Outlining the parties to the agreement.
2. The definition of what is confidential, *i.e.*, the information be held confidential. Modern confidential agreements will typically include a list of types of items which are covered, including unpublished patent applications, know- how, scheme of financial information, verbal representations, customer lists, vendor lists, business practices (strategies etc.)
3. The disclosure period: information not disclosed during the period (*e.g.*, one year after the date of NDA) is not deemed confidential.
4. The exclusions from confidential information: use of the confidential data will be invalid if

i) The recipient had prior knowledge of the materials.

ii) The recipient gained subsequent knowledge of the materials from another source.

iii) The materials are generally available to the public.

iv) Is disclosed by receiving party with disclosing party's prior written approval.

5. Provisions restricting the transfer of data in violation of national security.
6. The term (in years) of the confidentiality.
7. The term (in years) the agreement is binding.
8. Permission to obtain ex- parte injunctive relief.
9. The obligations of the recipient regarding the confidential information: include some version of obligations:

   i) To use the information only for enumerated purposes.

   ii) To disclose it only to persons with a need to know the information for these purposes.

   iii) To use appropriate efforts to keep the information secure. Reasonable efforts are defined as a standard of care relating to confidential information that is no less rigorous than that which the recipient uses to keep its own similar information secure.

   iv) To ensure that anyone to whom the information is disclosed further abides by obligations restricting disclosure and ensuring security at least as protective as the agreement.

10. Types of permissible disclosures: such as those required by law or court order. If a court finds any provision of this agreement is invalid or unenforceable, the remainder of this

agreement shall be interpreted so as best to effect the intent of the parties.

11. The law and jurisdiction governing the parties. The parties may choose exclusive jurisdiction of a court of a country.
12. Waiver: the failure to exercise any right provided in the agreement shall not be a waiver of prior or subsequent rights.

The agreement and each party's obligations shall be binding on the representatives, assigns and successors of such party. Each party has signed this agreement through its authorized representatives.

12

Writing PhD Synopsis and PhD Thesis

PhD SYNOPSIS / PROPOSAL FOR REGISTRATION

A research proposal or PhD Synopsis for Ph.D. registration, whether the area of study belongs to natural sciences, social sciences, languages, medicine or engineering, is necessary to be prepared and submitted to the authorities for acceptance. This is the first document of a PhD program and any person makes up his mind about the project based on this document only. Every University has its own procedure of writing and submitting the PhD proposal to maintain the standards yet, certain basic components, in which a number of questions need to be addressed while preparing a PhD synopsis. Why research on the proposed topic should be undertaken and what gains are likely to be achieved? What has been done previously in this or related areas? What are the objectives of this study and how these will be achieved? Are the facilities required for doing the proposed research available? An extensive initial exercise should help in designing a sound research project, which is likely to make a significant contribution in successful completion of Ph.D. research.

COMPONENTS OF A SYNOPSIS

The following components should be provided in a synopsis of a PhD research project. The details may, however, vary according to the field of study. Any alteration to the following format may be made in a specific discipline only with good justification.

Title Page

A title page of the synopsis should include title of the research project, name of the student (with qualifications), name of the supervisor(s), place of work and date (month and year) of submission.

Topic

The topic for research should be selected carefully. It should be specific and worded to show the nature of work involved as far as possible. This should be brief and self-explanatory. It should relate directly to the main objective of the proposed research. A more specific and descriptive sub-title can be added if necessary, for example to indicate the main methodology that will be applied.

Introduction

It should provide a brief description to introduce the area of the proposed research work. In this, one should introduce the main problem, set it into context and introduce the particular niche within the main subject area that will be worked with. For example, the main subject area could be deforestation and the Introduction would then briefly argue why it is relevant to be concerned with deforestation – to whom it is a problem and why.

Review of Literature

A review of the relevant literature showing the work done previously in the area of proposed research is essential to plan further research effectively. The information given in the review should be supported by references. One should present documentation of the existence of the problem, how it is manifested, who it affects and involves, what roles and interests the involved actors have, the historical background to the problem. The problem analysis is based on a critical review of scientific literature: the theories typically used to frame research on the subject area, knowledge available and research methods used with what degree of success. The review can add to the justification of choice of the subject included in the Introduction. It is important that the review includes recent literature, and that it critically synthesises knowledge within the subject being addressed rather than merely describing it.

Justification and Likely Benefits

It is important to provide justification for undertaking the proposed research, perhaps in the light of previous work done. It should be possible in most cases to anticipate the specific and general benefits likely to be achieved as a result of completion of the proposed research.

Objectives

Broad objectives as visualized to be achieved should be clearly outlined and these should be itemized. The objectives will indicate the major aspects of the study to be undertaken. These should be identified on the basis of the problem analysis. The objectives should focus on concepts and problems mentioned in the problem analysis Each research proposal should contain one overall objective describing the general contribution that the research project makes to the subject area as well as one or more specific objectives focusing on discrete tasks that will be achieved during the research.

Plan of Work and Methodology

A plan of work describing the various aspects of the study in a logical sequence along with the methodologies to be employed, are the most important aspects of any research plan. Sufficient details to demonstrate that the researcher has a fairly good idea about the nature of work likely to be involved should be provided. In the case of experimental sciences, *e.g.*, which equipments and experimental procedures will be used to obtain the results; in the case of social sciences what resource materials will be used; whether the required information will be obtained from primary or secondary sources, etc. A time schedule for the various aspects of the proposed research may be provided wherever possible.

Place of Work and Facilities Available

In order to complete the proposed research, some specialized facilities may be required. For example in case of experimental sciences different equipments may be involved or in the case of, may be, a study on a scholar, the relevant literature may be available in a foreign country. Therefore it is important to identify the place where the research work will be undertaken and whether the resources and facilities required for doing the research are available.

References and Bibliography

Synopsis should contain at the end a list of references, and a bibliography if required. These should be written on a standard pattern.

It will be difficult to define an overall length for a synopsis for Ph.D. research in such varied fields of study. Whereas it should be concise as far as possible and avoid repetitions, it should also provide sufficient details on the various aspects mentioned above to show that the research involved has been well understood and planned, and it is of an acceptable academic merit. The total length of a synopsis may run from 1,500 to a few thousand words. General guidelines are presented to prepare a PhD synopsis.

PhD SYNOPSIS FOR SUBMISSION

The term PhD synopsis is also considered as a detailed summary of the work with important results highlighting the original contributions in the thesis to be submitted. It should give an outline of the thesis. The review of earlier work is to be minimized with just enough to highlight the contributions in the research work to be reported in the thesis. It is expected that at the time of submission of the synopsis no work is yet to be completed except writing the thesis and all other academic requirements such as course work, comprehensive examinations and the suggestions and directions given by members of the Doctoral Committee have been fulfilled.

LAY OUT OF SYNOPSIS

- Cover Page & Title page
- Declaration
- Body of the Synopsis
  - Motivation and Problem statement
  - Brief survey of earlier work
  - Overview of the thesis
  - Major contributions (if required with results/ graphs/ photographs)
  - Conclusions
- List of References (pointed references only in the body)
- List of Publications (from the PhD work)

Cover Page & Title Page

Every University has a set pattern of cover page or title page. Mainly it contains Title of the thesis, course name, specialization, name of the student, name of the supervisor, year and name of the department and University.

Declaration

The declaration is required in every University and the declaration is made regarding originality of the work and also that the work is carried by a particular person under the supervision of assigned guide and the work is not considered for any other degree.

Motivation and Problem Statement

It has brief detail of the problem and rationale of the project. The objectives of the project are also discussed here.

Brief Survey of Earlier Work

It contains the detail of earlier work done and literature survey.

Overview of the Thesis/ Major Contributions

This is the major section where detail of the project is made. The methodology adopted and results obtained are discussed in this section. The major deliverables, achievements and contributions of the project are discussed in this section. If required the results are supported with graphs and photographs.

Conclusions

The main conclusion of the work is written here.

List of References

The references pointed in the body are presented here.

List of Publications

The publications made out of the project are written here. Accepted papers are also mentioned in this section. The details of presentations made in seminars, conferences and conventions out of the project work can also be written here.

PhD THESIS

A PhD thesis is a research report submitted by a researcher who has carried out research on an assigned topic of research under the supervision of a supervisor/guide to obtain the degree of doctor of philosophy. The report concerns a problem or series of problems in the area of a research carried out by a researcher and it should describe what was known about it previously, what was done towards solving it, what is thought about the results mean, and where or how further progress in the field can be made. A particular thesis will also be used as a scientific report and consulted by future workers in the concerned laboratory who will want to know, in detail, what was carried out on particular topic. Theses are occasionally consulted by people from other institutions, and the library also.

The text written in thesis must be clear. Good grammar and thoughtful writing will make the thesis easier to read. Scientific writing has to be a little formal—more formal than this text. Native English speakers should remember that scientific English is an international language. Slang and informal writing will be harder for a non-native speaker to understand. Sometimes it is easier to present information and arguments as a series of numbered points, rather than as one or more long and awkward paragraphs. A list of points is usually easier to write. One should be careful not to use this presentation too much: the thesis must be a connected, convincing argument, not just a list of facts and observations.

A typical thesis has following steps:

- Title Page
- Abstract / Summary
- Acknowledgements
- Abbreviations
- Contents
- List of Tables (where applicable)
- List of Figures (where applicable)
- Introduction (including literature review) or
  - Introduction
  - Review of Literature

  As separate chapters as per requisite of the subject
- Material and Methods
- Results

  May comprised of one chapter or a number of chapters depending upon the subject matter/ requirements
- Discussion (including Conclusion/s, Recommendation/s where applicable)
- References / Bibliography / Literature Cited
- Appendices (where applicable)
- Any other information specific to the respective discipline

PATTERN OF A PHD THESIS

All theses presented in typescript for the degree of Ph.D should normally comply with the following specifications unless specified or permission to do otherwise is obtained from the relevant authority / body.

Size of Paper

A4 size be used, no restriction is placed on drawings and maps.

Paper Specification

Usually Six copies on good quality paper (minimum 80 gsm) are submitted.

Method of Production

The text must be typewritten in acceptable type face and the original typescript (or copy of equal quality) must normally be submitted as the first copy. The second and subsequent copies may be produced by means of other acceptable copying methods.

Layout of Script

Typescript should appear on one side only, lines; at least one-and-a-half spaced. Footnotes, quotations, references and photographic captions may be single spaced. Where appropriate, these should contain lists giving the locations of figures and illustrations.

Font Size

Title Page	18-22
Headings / subheadings	14-20

Text	10-12
Footnotes	8-10

Footnotes are given on the same page where reference is quoted.

Type Style

Times New Roman / Arial / Courier New / Univers

Margins

At least 1¼ -1½ inches (3.17-3.81 cm) on the left-hand side, 3/4 - 1 inch (2 -2.54 cm) at the top and bottom of the page, and about ½ - 0.75 inches (1.27 - 1.90 cm) at the outer edge should be kept. The best position for the page number is at top-centre or top right ½ inches (1.27 cm) below the edge. Pages containing figures and illustration should be suitable paginated.

Following is the preferable layout of the thesis

Title page

All theses must contain a title page giving the title of the thesis, the author's name, the name of the degree for which it is presented, the department in which the author has worked or the Faculty to which the work is being presented, and the month and year of submission. A typical specimen is presented below in Fig. 12.1.

Length of thesis

Whilst most of the regulations do not contain a clause relating to the maximum length of theses, it is expected that work presented for the degree of Ph.D. should normally between 40,000–120,000 words of text. Candidates wishing to greatly exceed these sizes should discuss the matter with their supervisors.

Publish work

Published work from the theses is included as appendix (Reprints/ proof / preprint).

Binding

All final theses and published work presented for higher degrees must be bound in a permanent form or sometimes in a temporary (hard binding are provided after defense of the thesis) form approved by the Advanced Studies and Research Board; where printed pamphlets or off-prints are submitted in support of a thesis, they must be bound in with the thesis, or bound in such manner as Binderies may advise. Front cover should give title of the thesis, name of the candidate and the name of the Institute/ Department/ Centre/ College through which submitted, in the same order from top to bottom. The lettering may be in boldface and properly spaced. Their sizes should be: title 24 pt., name of the candidate 18 pt. and the name of the department/ institute/ centre/ college 18 pt.

<Font Style Times New Roman - Bold>

STUDIES ON SOME MEDICINALLY IMPORTANT FLAVONOIDS

Font Size 18> <1.5 line spacing>

THESIS

<Font Size 14>

Submitted by

<Font Size 14> <Italic>

A. BHARGAVA

<Font Size 16>

In Partial Fulfillment of the Requirements

For the Degree of

<Font Size 14> <1.5 line spacing>

DOCTOR OF PHILOSOPHY

<Font Size 16>

DEPARTMENT OF PHARMACEUTICAL SCIENCES

MAHARSHI DAYANAND UNIVERSITY,

ROHTAK-124001, HARYANA

<Font Size 16><l.5 line spacing>

DECEMBER 2014

<Font Size 14>

Fig. 12.1: A typical specimen of title page & cover page

13

Use of MS Word, MS PowerPoint and MS Excel

MICROSOFT WORD 2010

First of all to get started with Microsoft Word, you need to locate and open the program on computer to access the Microsoft word from the Start Menu (Fig. 13.1). Click on the button in the bottom left corner on Start Menu. You will see the MS Word icon and by clicking on the Microsoft word option, you can open Microsoft word window, if you do not find the Microsoft word from here then click on "All Programs" and search the Microsoft word from there and click on Microsoft word.

After left click on the Microsoft word the window will be open, as given in Fig. 13.2.

This box gives two important informations: the name of the file that you are currently working on (in this case, "Document 2" since the document is not yet renamed) and which program you are using ("Microsoft Word").

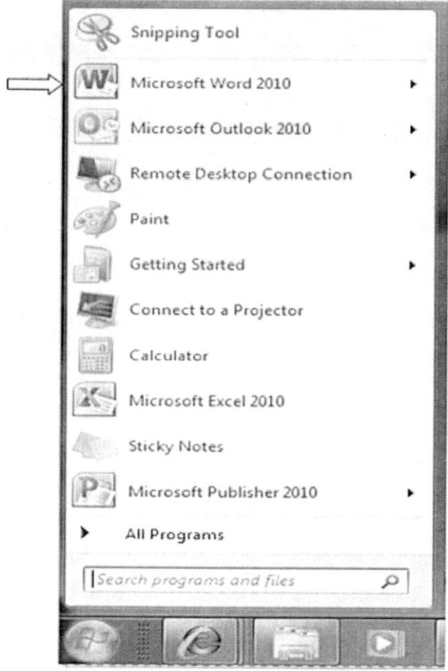

Fig. 13.1: How to open microsoft word from start menu

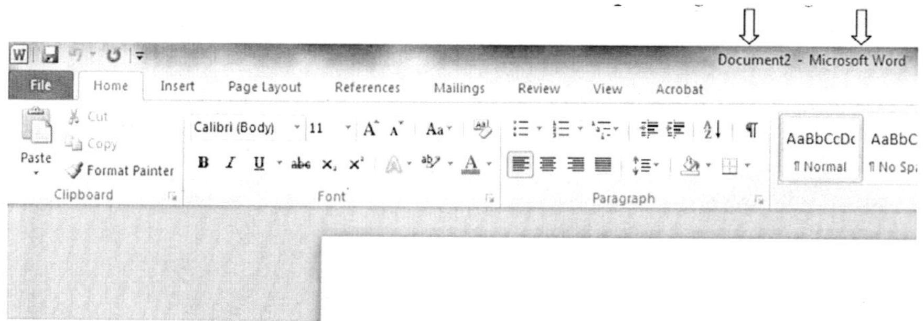

Fig. 13.2: Microsoft word window

Microsoft Word Features

1. **Title bar:** This is the close view of the title bar (Fig. 13.3) which shows the name of the document, here, it is "Document 2.

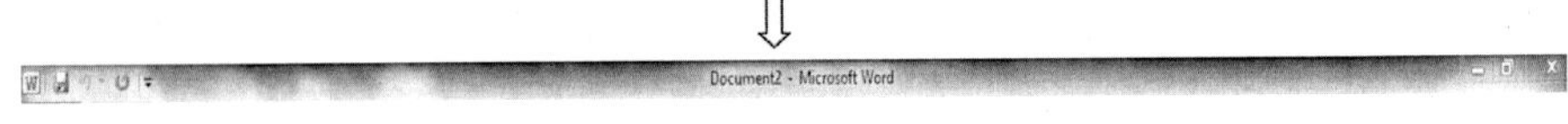

Fig. 13.3: Close view of the title bar

***Minimize*:** Left-click this button to shrink the window to a small button that will appear in the task bar on right side.

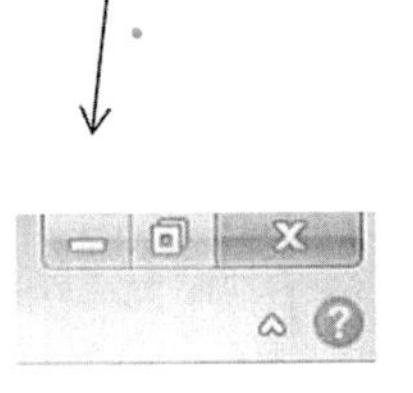

***Maximize and Restore*:** Left-click this button to make fit on the screen and with again left click on this button window will appear in its original size.

***Close*:** By the left-click on this button you can close the window, document/ program. First of all make sure that your work is saved or not, before closing.

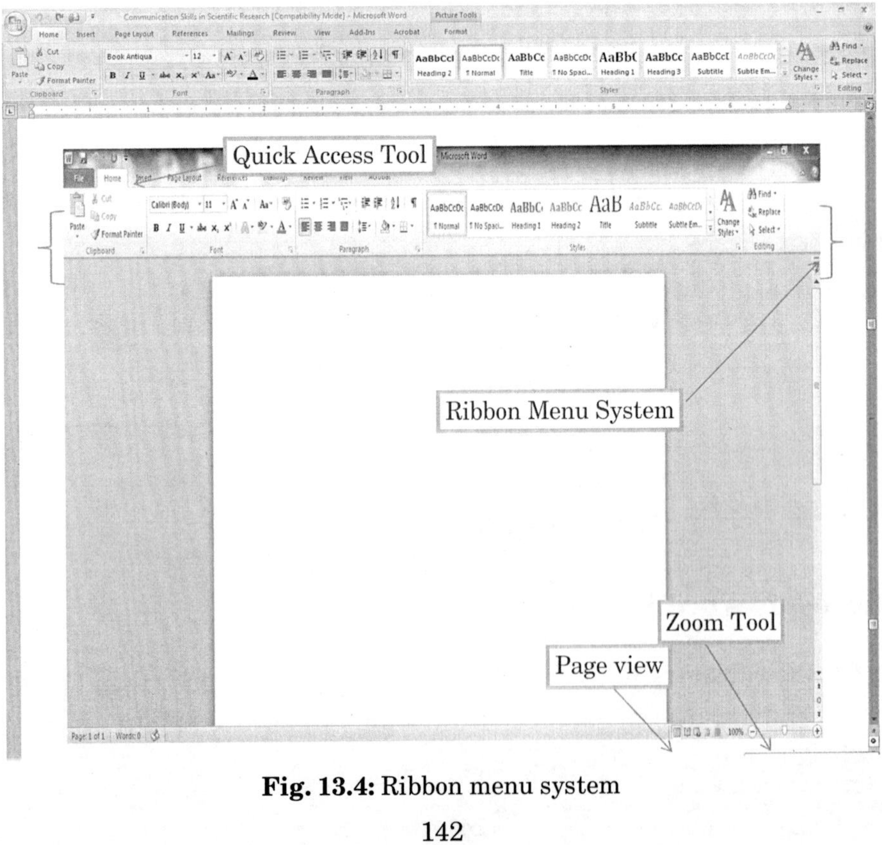

Fig. 13.4: Ribbon menu system

2. ***Ribbon menu system:*** In the ribbon menu system (Fig. 13.4) you can use all the tools of the word directly without going in every menu. Only by clicking on the menu all the function of the menu will come in the ribbon menu. In word 2010 all the functions are given in the ribbon menu system and these are more in function and numbers as compared to the Microsoft word 2007.

The File Menu

In Microsoft Office 2007, there was Microsoft Office logo Button on the top left-hand corner. In Microsoft Office 2010, this logo has been replaced by the Ribbon called "File" (Fig. 13.5). When you left-click on this, a drop-down menu appears. From this menu, you can perform function such as: Create a new document, open existing files; save files in a variety of ways and in different types of files in word mode, and print and exit commands are also present here.

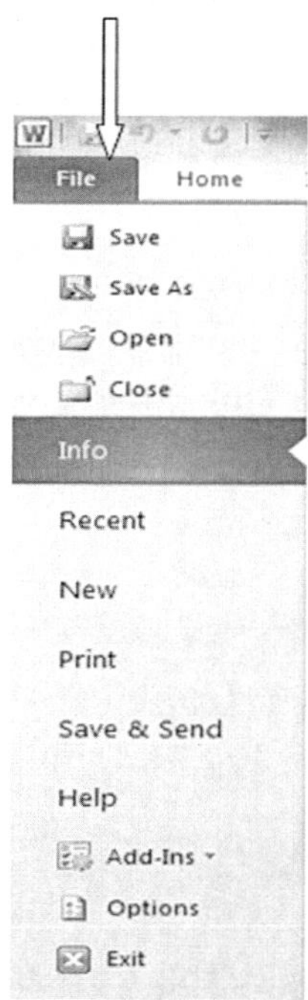

Fig. 13.5: The file menu

Home Tab

In the home tab different function are present like the cut, copy, paste and numbering systems etc., which you can apply in your Microsoft document. This is shown below in Fig. 13.6.

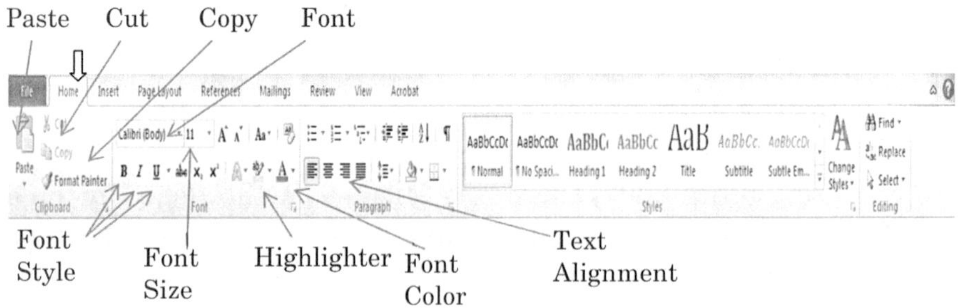

Fig. 13.6: Home tab window

Using the home tab you can make the document in which style or format you want like you can use Times New Roman and 12 font size for making the word document. In the above figure Calibri body indicate the font and 11 indicates the size of your font.

The Ruler

The ruler (Fig. 13.7) is present below the Ribbon. The ruler shows the actual dimensions of the page, by the use of the ruler you can set margins of the document. The ruler is entirely customizable and you can create a document of dimensions according to your needs. The default setting is 8.5 × 11 inches.

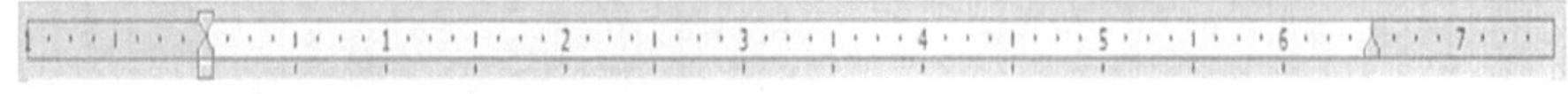

Fig. 13.7: Ruler

Insert Menu

In the insert menu (Fig. 13.8) there are different commands available like cover page, blank page, header footer, page no. option, shapes, charts which are impotent for the research purpose.

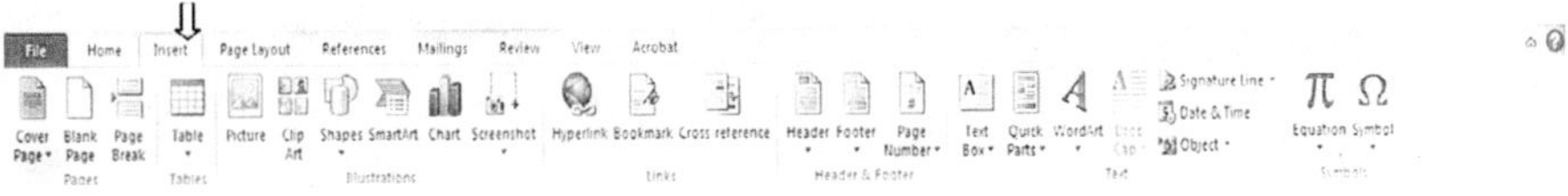

Fig. 13.8: Insert menu window

Cover Page

The cover page in which you can make the cover page for the document in which title of the research work and the name of supervisor and name of the candidate and the name of the university is present.

Table

In this menu you can insert the table in your document to depict your results of the experiments and the surveys on the basis of which you can make conclusion. You can also insert column and rows by putting the no. of the rows and the column.

Picture

In this menu you can insert the picture of the instrument which is used in your research work.

Shape, Smart Art and Charts

In this menu you can insert the different shapes like circles, squares, arrow, simple line boxes and charts by using different values. Fig. 13.9 shows making of chart in the word document using different values.

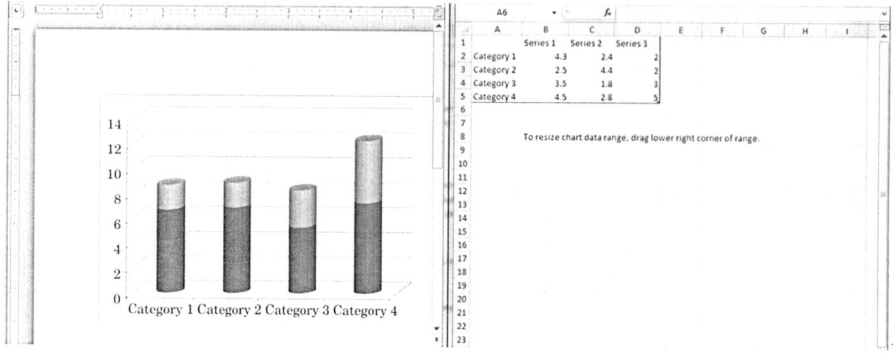

Fig. 13.9: Making of chart in the word document using different values

Making of Chart by Putting Values in the Microsoft Word

In case of the smart art different types of the charts are given which are editable in which you can put our information regarding the research. Fig. 13.10 shows different types of the smart art graphics.

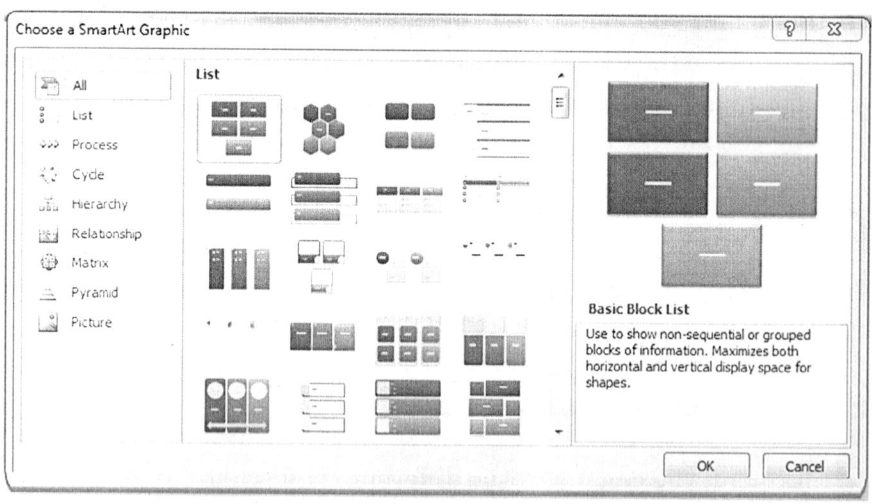

Fig. 13.10: Different types of the smart art graphics

Screen Shot

In this menu you can insert the screen shot of your computer window or the page which is currently opened.

Hyperlinks, Bookmarks and Cross Reference

You can set the hyperlink in your document by using this command *i.e.*, you can link to other document. By clicking on hyperlink, you can go directly on that file or on the information.

Bookmark is created in a document to give a name to specific point in the document. In case of the cross reference, by clicking on the cross reference you can go on the table or on the specific location in the word document.

Header, Footer and Page Number

Header, footer and page number is used to insert date, time, title of the chapter and title of the thesis or title of the other valuable documents on the page and the page number.

Text Box and Art Word

Text box command is used mainly in case of the newspaper formatting. Here by the use of this command the data are fitted in to the boxes. By the help of this you can provide more data in less space. In art word there are different types of styles of letters by which you can make an effective and attractive word document.

Symbols and Equations

By symbols and equation you can insert different symbols in your word document. A huge variety of the symbols and equation like binomial theorem, area of circle and expansion of sum are given in this menu.

Page Layout Menu

In the page layout menu (Fig. 13.11) all setting related to the page like the margin of the page and water mark on the page, orientation of the page, page border, page color, page column and page break is present. The figure of the page layout menu is given below.

Fig. 13.11: Page layout menu window

Reference Menu

In the reference menu you can set your references according to choice by putting the citation to the references menu.

Method of Reference setting in word:

1. First click References tab
2. Then click Manage Sources on the Citations & Bibliography menu
3. After that create new sources that will automatically be added to both the Current and Master List
4. Sources in the Current List will be shown in dropdown Insert Citation list. Make your selection.
5. Enter information about each source.
6. After entering all your sources, close the window.
7. Select the Style on the Citations & Bibliography menu and choose the appropriate style
8. Click on Bibliography dropdown the list and select Insert Bibliography

9. The bibliography will appear in Word doc.
10. Edit accordingly

The Fig. 13.12 shows the reference menu.

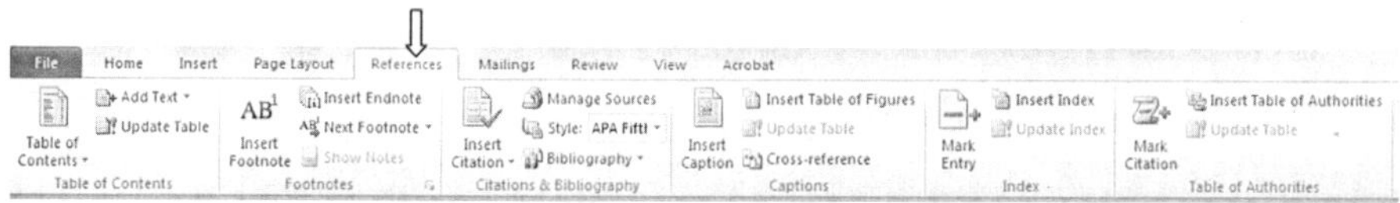

Fig. 13.12: Reference menu window

Mailing Menu

Mailing menu is very impotent menu from the communication point of view. Different design of the envelope and the label are given here in the menu. By the mail merge you can merge the two documents like envelope and the main letter and can send it to the people. This menu can be started by starting mail merge in which letter, envelope label and directory are present. In the select recipients menu you can select the recipients from the existing list or can create a new list. The mailing menu is shown in Fig. 13.13.

Fig. 13.13: Mailing menu window

Review Menu

Review menu (Fig. 13.14) is very important menu. In this menu, you can check the grammar and spelling of the word documents by left clicking on the icon.

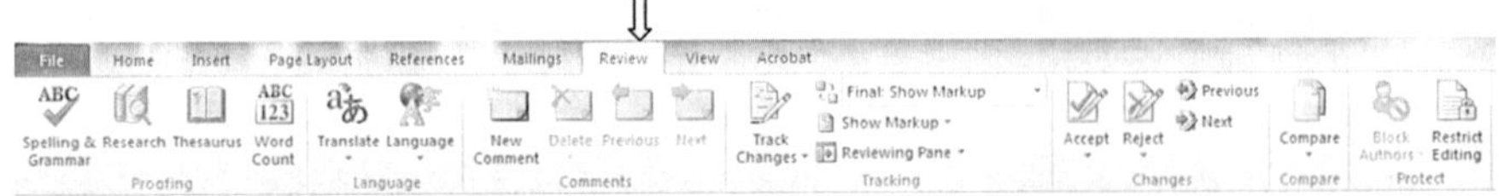

Fig. 13.14: Review menu window

Spelling and Grammar

Spelling and grammar icon is used for checking the spelling and the grammatical mistakes in the document. The incorrect word in the document appears in red color underline, is not available in the dictionary box.

In suggestion box, you will see the suggestions for replacing the word which has been misspell. You can select anyone of these buttons. Then click over "change button" the word will be replaced by the selected one. You can click "ignore" button if the change is not required. You can click "ignore all" if you want to continue without changes in the word document. Click "add" button to add the word in the dictionary, Click the "ok" button.

Word Count

Microsoft word provides the facility of counting the words by selecting the text whose words are to be counted. The word count tells the total complete statistics of numbers of the pages, word and characters.

Thesaurus

Using thesaurus you can change a word with any of its synonyms. In this way, you can prevent the repetitive use of the single word which shall bring beauty in the word document.

Language

In the language menu two icons are present. In "translate" icon, first select the word which you want to translate in the other language. Then click to the "translate" option and then select the language in which the word have to be changed, then click on "ok" button. The document will be changed in the required language.

Comments

New comments, delete comments, previous and next comments are present in this menu. By clicking left click on the word where you want to give the comment in the document you can insert the comment. You can delete the comment by using delete comment icon; you can go from one comment to the next comment by using the previous comment and next comments.

Compare

By using this icon, you can compare the two word files for matching the correct and the other files.

Restrict Editing

This icon is used for making the word file secure and safe for editing so that nobody can change the word file.

View Menu

Print Layout

In this menu by clicking on the print layout icon you can see the print overview of the page. In this you can notify changes which you have to do.

Full Screen Reading

By clicking on this icon you can read the word document like a book.

Web Layout

In this you can see the document as a lay out on the web. In which you zoom the word document. The Fig. 13.15 shows the view menu.

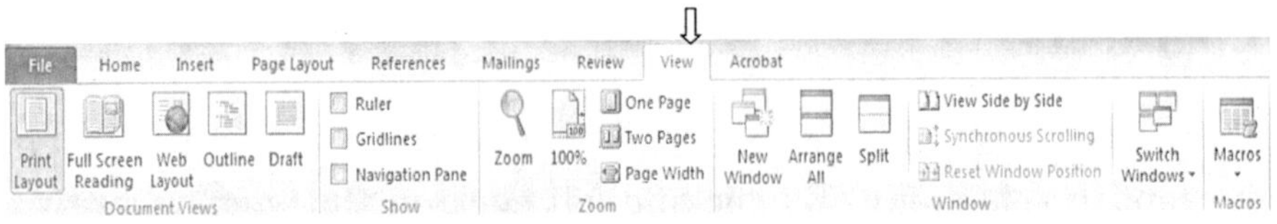

Fig. 13.15: View menu window

MICROSOFT EXCEL 2010

Microsoft excel is spreadsheet program under window. It is good program for calculations, recalculation and repetitive information. In the Microsoft excel, multiple copies of the worksheet can be made, and frequent change can be done in the Microsoft excel program. It is very important application for the business point of view *i.e.*, inventory control, stock investments, budgetary control and what if analysis. A Fig. 13.16 shows Microsoft excel window.

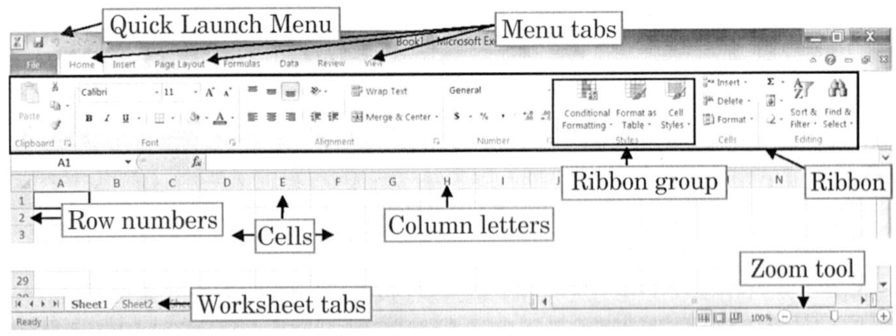

Fig. 13.16: Microsoft excel window

Entering and Editing Data

For entering the data in excel first of all click on the cell with the help of mouse. For changing the content of the cell of the excel sheet, press delete key or click in the formula bar to edit in the content.

To Move Around a Spreadsheet

You can move one cell down with the help of the enter key and with the help of the shift + Enter key you can move up side through the cells. Similarly you can move through the right side by using the tab key. Use shift + Tab key to go to left side through the cells.

Change the Width of a Column

You can change the width of a column (Fig. 13.17) by using mouse pointer, take the pointer on the right edge of the cell and press the left click and hold it up to width you want and through this you can adjust the width of the cell according to your requirement.

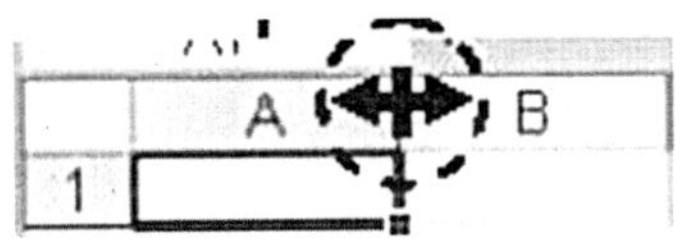

Fig. 13.17: Width of a column

Change the Color or Style of Text

The Fig. 13.18 shows how to change the color and text style.

Alignment

Alignment of the text along the left of the cell, along the right of the cell, or in the center of the cell can be done with the help of this menu (Fig. 13.19).

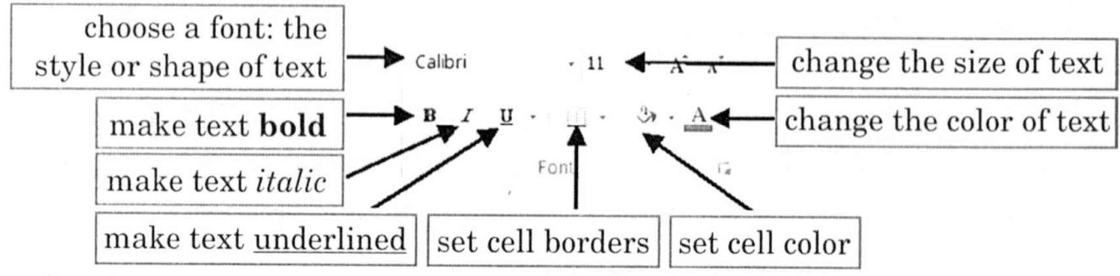

Fig. 13.18: How to change the color and text style

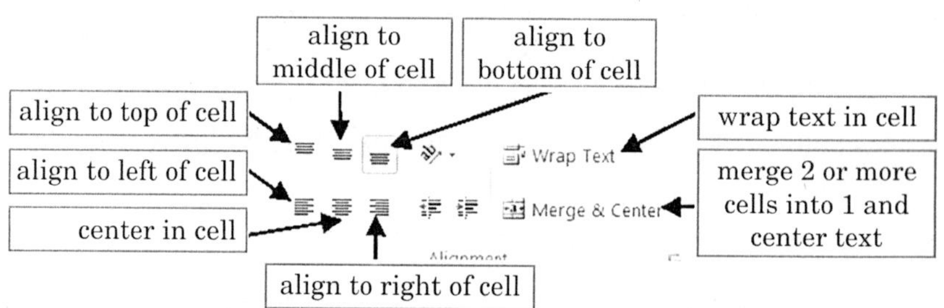

Fig. 13.19: How to change the alignment

Number Menu

In the number menu (Fig. 13.20) you can insert the numbers and decimal in the cell, you can also increase or decrease the decimal places. You can choose the format in this menu like the general format and format as dollar value.

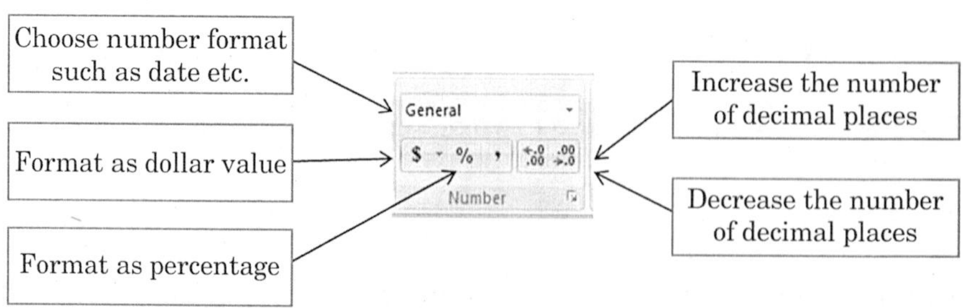

Fig. 13.20: How to insert the numbers and decimals from number menu

Style Menu

In this menu different types of the colored formats of the tables and the cells are given, you can format the cell and table according to your needs. Suppose you want the different color of the upper row cell and the lower rows cells, with the help of this menu you can format the cells according to your needs.

Cell Menu

In this menu, you can insert or delete the cell in the worksheet and format the sheet according to need. You can also rename the worksheet; can change the row height, row width and protection of the worksheet.

Edit Menu

With the help of the edit menu you can edit the cells by erasing or by changing the values of the cell and can apply the command of auto sum, average count number etc.

Insert Menu

In this menu you can insert different types of the graphs, picture, clip art, shapes of different types, column line, pie chart, bar diagram, hyper link, header and footer etc. The Fig. 13.21 shows the insert menu of the Microsoft excel.

Fig. 13.21: Insert menu of the microsoft excel window

Creating Chart in Microsoft Excel

First, select the data which have to be included in the chart by clicking and dragging with the mouse pointer. Then:

Insert > choose style from Charts ribbon.

Next, choose options on each box and click Next or Finish.

Page Layout Menu

In the page layout menu all setting related to the page like the margin of the page and, orientation of the page, page border and page color, page column and page break is present. The Fig. 13.22 shows the page layout menu window.

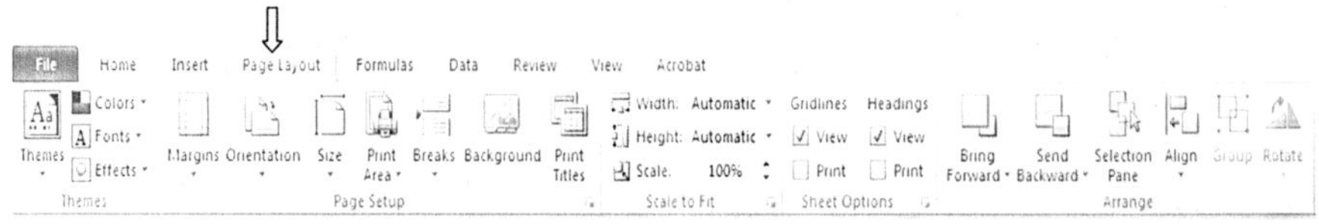

Fig. 13.22: Page layout menu window

Formula Menu

Formula menu of the Microsoft excel contains the various mathematical formulas like the auto sum, average, trigonometric formulas, financial and statistical formulas. You can apply these formulas in your business, research work etc. The Fig. 13.23 shows the formula menu window.

Fig. 13.23: Formula menu window

Data Menu

In the data menu you can access the data from the computer and the internet source, or from the existing source like sharing with the other computer. In this menu you can validate the data and can remove the duplicate data. The Fig. 13.24 shows the data menu window.

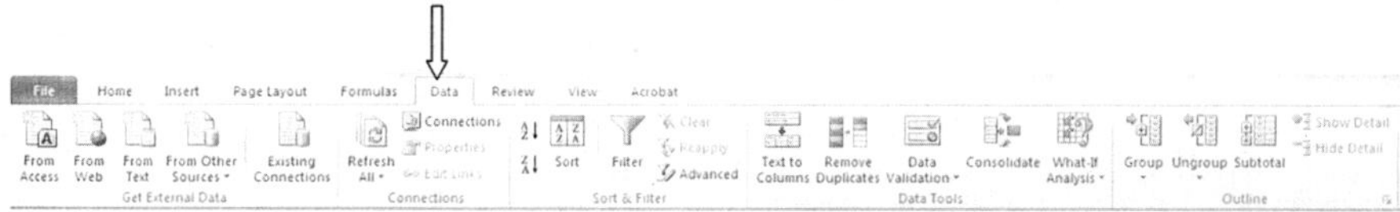

Fig. 13.24: Data menu window

Review Menu

Review menu (Fig. 13.25) is very important menu. In this menu you can check the grammar and spelling of the word documents by left clicking on the icon.

Fig. 13.25: Review menu window

Spelling

Spelling and grammar icon is used for checking the spelling mistakes in the document. The incorrect word appear in red color underline is not available in the dictionary box.

In suggestion box, you will see the suggestions for replacing the word which has been misspell. You can select anyone of these buttons.

Then click over "change" button the word will be replaced by the selected one. You can click "ignore" button if the change is not required. You can click "ignore" all if you want to continue without changes in the word. Click "add" button to add the word in the dictionary, Click the "ok" button.

Thesaurus

Using thesaurus you can change a word with any of its synonyms. In this way you can prevent the repetitive use of the single word which shall bring beauty to the document.

Language

In the language menu two icons are present. In translate icon, first select the word which you want to translate in the other language. Then click on the translate option and then select the language in which the word have to be changed then click ok button. The document will be changed in the required language.

Comments

New comments delete comments previous and next comments are present in this menu. By clicking left click on the word or cell, where you want to give the comment in the document you can insert the comment. You can delete the comment by using delete comment icon; you can go from one comment to the next comment by using the previous comment and next comments.

Protect Sheet and Protect Workbook

This icon is used for making the word file secure and safe for editing so that anyone cannot change the file.

View Menu

In the view menu you can view the page layout, page break view,

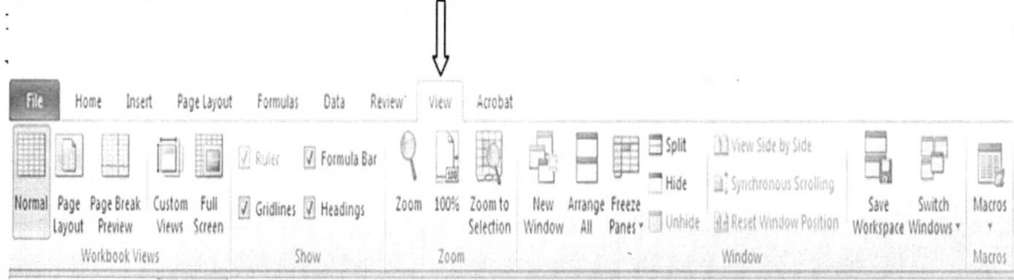

Fig. 13.26: View menu window

MICROSOFT POWERPOINT 2010

PowerPoint is a system in the Microsoft Office that enables us to present information in office meetings, seminars and lectures to create maximum impact in a minimum time. With the help of the Microsoft power point you can pass your information very quickly and can make evaluation fast.

For accessing the Microsoft power point from the Start Menu, Click on the button in the bottom left corner on Start Menu. You will see the MS Power Point icon here and by clicking on the Microsoft power point you can open Microsoft word window, if you do not find the Microsoft power point from here then click on "All Programs" and search the Microsoft power point from there and click on Microsoft power point. Fig. 13.27 and Fig. 13.28 shows the Microsoft power point window.

This box gives two important informations: the name of the file that you are currently working on (in this case, "Presentation 1" since your presentation not yet renamed) and which program you are using ("Microsoft PowerPoint").

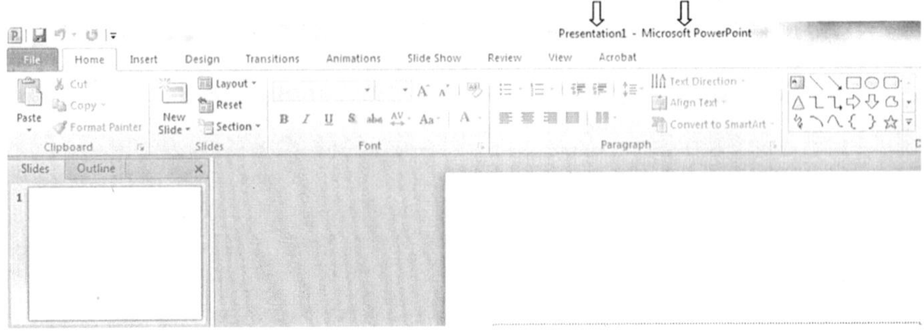

Fig. 13.27: Microsoft power point window

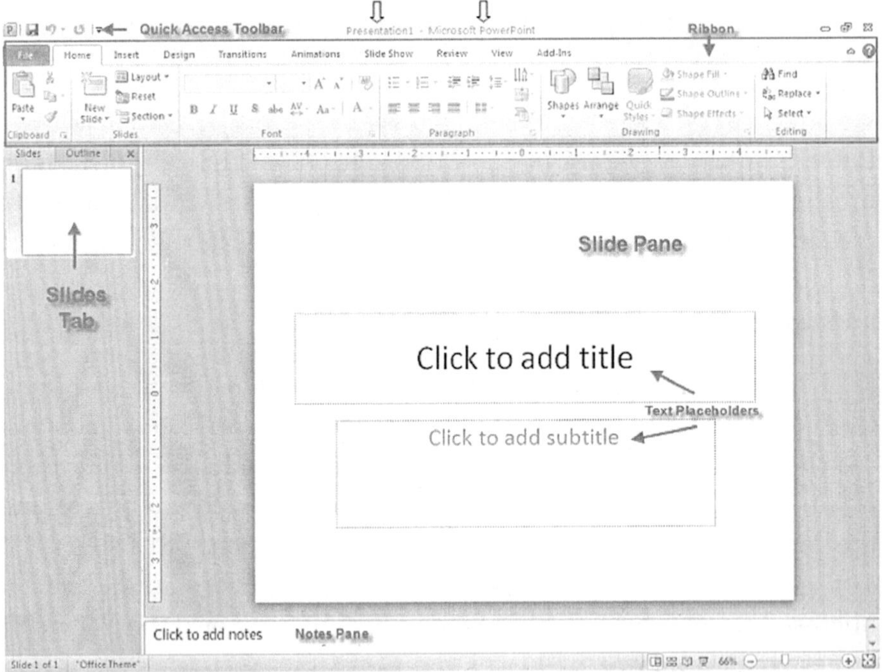

Fig. 13.28: Microsoft power point window

Quick Access Tool Bar

In the open access tool bar, save icon, undo and redo icon is present. This toolbar provides access to common functions such as Save or Undo Last Action.

Minimize: Left-click on this button will shrink the window to a small button that will appear in the task bar on right side.

Maximize and Restore: Left-click this button to make fit on the screen and with again left click on this button window will appear in its original size.

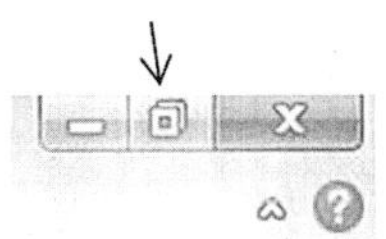

Close: By the left-click on this button you can close the window document/ program. First of all make sure that our work is saved or not before closing the window.

Home Menu

In the home menu of the power point presentation (Fig. 13.29) cut, copy, paste, lay out of the slide, language options, size of the letter, symbols are present. Home menu of the Microsoft power point is slightly different as compare to the Microsoft excel and Microsoft word.

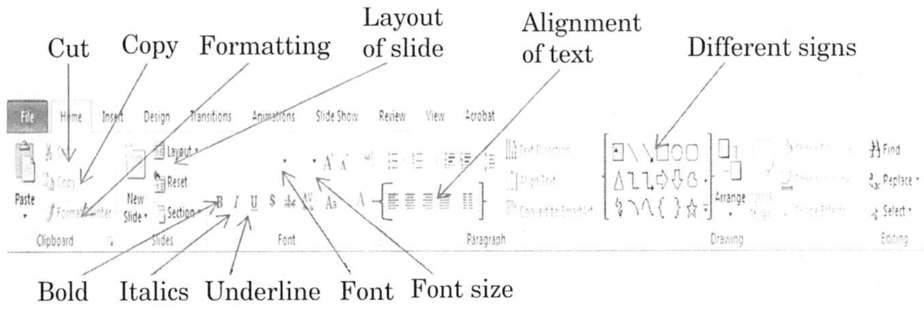

Fig. 13.29: Home menu of the power point presentation window

Insert Menu

You can insert table, picture, clipart, screen shot, photo album, shapes, smart art, chart, hyperlink, header, footer, date, time, equation, symbols, video and sounds. The Fig. 13.30 shows the insert menu window of power point presentation.

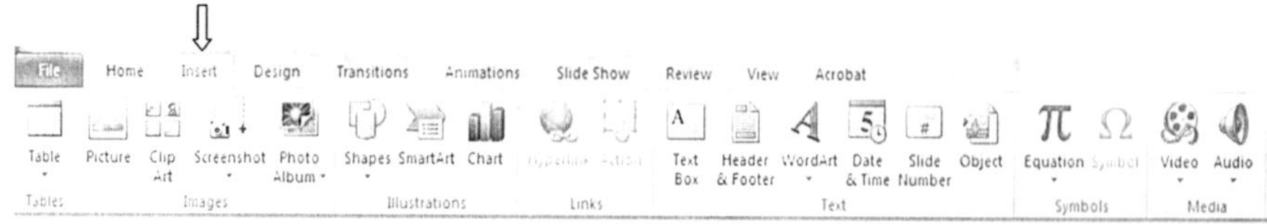

Fig. 13.30: Insert menu of the power point presentation window

Design Menu

In the design menu different types of the design of the slides are given, you can choose the design according to your requirement. In this menu, slide orientation and page setup is also present. The Fig. 13.31 shows the design menu window of power point presentation window.

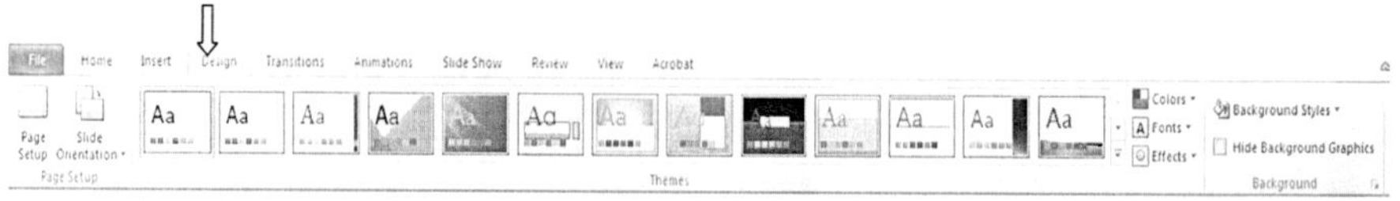

Fig. 13.31: Design menu of the power point presentation window

Transition and Animation Menu

In the transition menu different styles of the transition is given. The transition forms work during the slide show of your presentation. In this menu, you can choose separate menu transition for each slide and can set the time of the sound or video for the presentation. In the animation menu different styles of the animations are present. You can apply the animation to the slide by clicking on the icon of the animation. The Fig. 13.32 shows the design and animation menu window.

Slide Show

You can see your slide's actual presentation view with the help of the slide show menu. In the slide show menu you can view your slides from beginning or from the current slide by clicking on the icons given in the slide show menu. View of the slide show menu window is shown in Fig. 13.33.

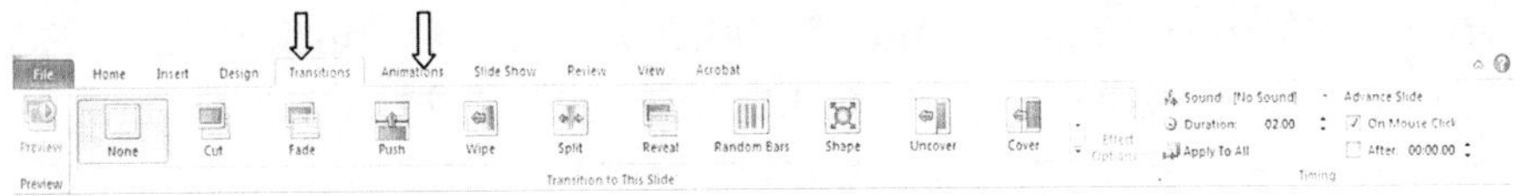

Fig. 13.32: Transition and animation menu of the power point presentation window

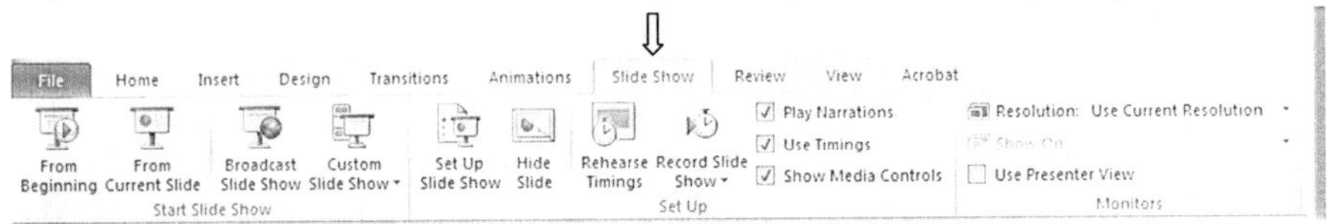

Fig. 13.33: View of the slide show menu window

Review Menu

Review menu (Fig. 13.34) is very important menu. In this menu you can check the spelling of the presentation by left clicking on the icon.

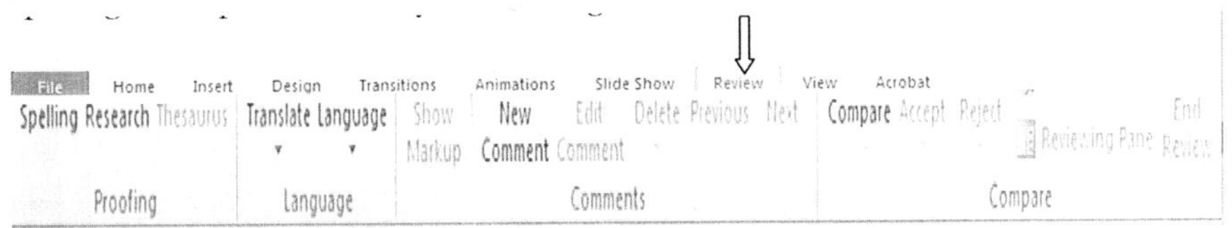

Fig. 13.34: Review menu window

Spelling

Spelling and grammar icon is used for checking the spelling mistakes in the document. The incorrect word appear in red color underline is not available in the dictionary box.

In suggestion box, you will see the suggestions for replacing the word which has been misspell. You can select anyone of these buttons. Then click over "change" button the word will be replaced by the selected one. You can click "ignore" button if the change is not required. You can click "ignore" all if you want to continue without changes in the word. Click "add" button to add the word in the dictionary, Click the "ok" button.

Thesaurus

Using thesaurus you can change a word with any of its synonyms. In this way you can prevent the repetitive use of the single word which shall bring beauty to the document.

Language

In the language menu two icons are present. In translate icon, first select the word which you want to translate in the other language. Then click on the translate option and then select the language in which the word have to be changed then click ok button. The document will be changed in the required language.

Comments

New comments delete comments previous and next comments are present in this menu. By clicking left click on the word or cell, where you want to give the comment in the document you can insert the comment. You can delete the comment by using delete comment icon; you can go from one comment to the next comment by using the previous comment and next comments.

Protect Sheet and Protect Workbook

This icon is used for making the word file secure and safe for editing so that anyone cannot change the file.

View Menu

In the view menu you can view the page layout, page break view, ruler, formula bar, zoom options. The Fig. 13.35 shows view menu window.

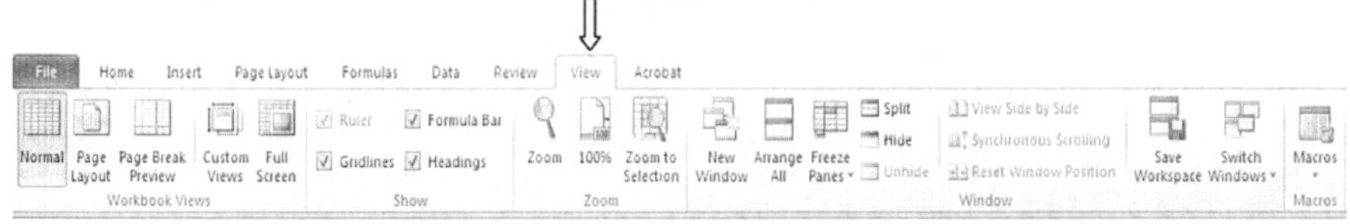

Fig. 13.35: View menu window

14

Concepts of Internet and Primary Search Engines in Research

INTERNET

Internet is a series of interconnected networks that allows the exchange of the data. It is the big connection of the cables and computer networks which are interconnected. Through which you can connect to world, can take the information which you want like the information related to the research, business and other field. There is no particular single agency which controls the internet in the world; different countries control the internet content according to their requirements.

USE OF INTERNET

There are various online softwares which are very useful in the field of research.

You can learn through distance learning program on the internet.

With the help of the internet you can access multimedia information like audio, video and images.

With the help of the internet you can connect with the other peoples to share information with the help of email or other source.

With the help of the internet, you can share the information and update any information very quickly.

PRIMARY SEARCH ENGINES IN RESEARCH

Google Books

With the help of the Google book search, you can search book of any topic of the research, various types of the books are present on the Google books.www.books.google.com

Google Scholar

Google scholar helps you to find out the abstract, papers and other scholar source with speed and accuracy.

Google Patent

Google patent is very useful in case of the patent; each Google patent search result shows information about each patent search very quickly. www.books.google.com/patent

RefSeek

RefSeek contains more than 1 billion documents, newspapers, books, youb pages, journals, and more, RefSeek is authoritative resource. www.refseek.com

Academic Index

This is the search engine only created for the college students. This search engine provides the information about the history, criminal justice, health and more. www.academicindex.net

Microsoft Academic Search

Over 38 million different publications can be found on Microsoft academic research. This search engine gives more and reliable information about the research.

Google Correlate

Google correlate search engine is search engine which correlate the research with the real world data.

The British Library Catalogues & Collections

The British library contains the catalogues, digital collections and printed data.

World Cat

World cat library contains items from 10,000 libraries worldwide, containing the books, CD, DVDs, and articles.

Biology Browser

This search engine is very useful for the researcher in the field of the Biology. It is a great resource for finding research, resources, and information in the field of biology. With the help of this search engine you can check out Zoological Record and BIOSIS Previews.http://www.biologybrowser.org/

Science.gov

This is USA government science portal, you can search more than 50 databases and 2,100 selected websites from 12 federal agencies with the help of this search engine. This contains millions of pages of U.S. government science information. This contains the information about agriculture, health, computer and communication etc.https://www.science.gov/

Wolfram Search Engine

This search engine is useful for the researcher in the field of the mathematics and in the filled of engineering.

Further Readings

1. Davis, M. *Scientific Papers and Presentations*, Academic Press, 1997.
2. Day, R.A. *How to write and publish a Scientific Paper*, 4th Ed., Cambridge University Press,1995.
3. Michaelson, H.B. *How to Write and Publish Engineering Papers and Reports*, 3rd Ed., Oryx Press, 1990.
4. O' Connor, M. *Writing Successfully in Science*, Academic Press, 1991.
5. Paradis, J.G. and Zimmerman, M.L. *The MIT Guide to Science and Engineering Communication*, MIT Press, 1997.
6. Anonymous (2003): Tips for conducting a literature review. Centre Alpha Plus. Available on http://alphaplus.ca/pdfs/litrev.pdf; accessed 12 November 2008.
7. Bem, D.J. (1995): Writing a review article for Psychological Bulletin. Psychological Bulletin 118 (2): 172-177.
8. Day, R.A., Gastel, B. (2006): How to write and publish a scientific paper. Sixth edition. Greenwood Press, Westport.
9. Noguchi, J. (2006): The science review article – An opportune genre in the construction of science. Linguistic Insights Volume 17. Peter Lang, Bern.

10. Ridley, D. (2008): The literature review – a step-by-step guide for students. Sage
11. Publications, London.
12. T. Rosenbaum, Effective Communication Skills, New York LTAP Center, 2005.
13. M. Derntl, Basics of Research Paper Writing and Publishing, 2009.
14. J. Peat, Scientific Writing- Easy When You Know How, London BMJ Books, 2002.
15. Gardner D. Plagiarism and How to Avoid It. 1999.
16. Clough P. Old and new challenges in automatic plagiarism detection. *Plagiarism Advisory Service 2003; 1-14.*
17. Scanlon PM, Neumann DR. Internet Plagiarism Among College Students *Journal of College Student Development 2002;* 43(3): 374-385
18. Groark M, Oblinger D, Choa M. Term Paper Mills, Anti-Plagiarism Tools and Academic Integrity. *EDUCAUSE 2001;* 40-48.
19. Bansal, P. "IPR Handbook for Pharma students and Researchers", 2007, Pharma Med Press, 47-71.

SAMPLE PROPOSAL – TRAVEL GRANT

UNIVERSITY GRANTS COMMISSION
BAHADUR SHAH ZAFAR MARG
NEW DELHI – 110 002

Application for getting financial assistance to attend international conferences / symposia under the 'Travel Grants' scheme for college teachers / College Librarians, Vice-chancellors, Commission Members and UGC Officers.

A. Details about the applicant

1.	Name	:	**Dr. xxxxxxx**
2.	Date of Birth (Age) (For college teachers/College Librarians the maximum age limit is up to superannuation on the date of participation in the conference) (For Vice-Chancellors, Commission Members and UGC Officers should be in service on the date of conference is held)	:	09-09-1974 (40 YEARS)
3.	Sex (Male/Female)	:	FEMALE
4.	Category SC/ST/OBC (excluding creamy layer) / General	:	GENERAL
5.	Designation/Basic Pay / Nature of Appointment (Whether Permanent/ Temporary)	:	PRINCIPAL / 37400-67000 AGP 10,000 / PERMANENT

6.	Official address with pin code Telephone	:	xxxxxxxxxxxxxxxxxxxxx
7.	Main Subject and Field of Specialization	:	Pharmaceutical Chemistry
8.	List of publications in the specific field (attach separate sheets)	:	Please see annexure-1
9.	Whether a member of national / International professional bodies	:	Yes
10.	If YES specify the name of the body (s)	:	Life member of Indian Pharmacy Graduates Association (IPGA) no.
11.	Name of the College where working and name of the University to which it is affiliated	:	xxxxxxxxxxxxxxxxxxxxxxxx
12.	Maximum age of superannuation in your college	:	62 years
13.	Date of superannuation	:	September 30, 2034

B. Conference details

14.	Name / title of the conference to be Attended	:	International conference on Advances in Plant Sciences, Chiang Mai, Thailand
15.	Name of the organizers with complete Address	:	
16.	Name of the country and town where the conference will be held	:	Chiang Mai, Thailand

17.	Duration of the conference (date, month & year)	:	November 14-18, 2012
18.	The role of the applicant in the conference/symposium		
(a)	Presiding/chairing a Session (if yes, attach documentary evidence)		
(b)	Delivering a plenary lecture/ invited talk (attach documentary evidence along with a copy of the full text of the lecture/talk)		
(c)	Presenting a paper (please attach abstract and full paper)	:	Presenting a research paper, copy of the abstract, acceptance letter and full paper is attached.
19.	Whether the paper has been accepted for presentation? (attach documentary evidence and a copy of the full paper to be presented in the conference)	:	Copy of the acceptance letter and full length paper is attached.
20.	Indicate the mode of presentation (attach documentary evidence) oral/poster / both	:	ORAL (proof attached)
21.	Indicate whether the paper has been coauthored. In case it is co-authored give names of the authors along with their addresses	:	Dr. Munish Garg, Department of Pharmaceutical sciences. M. D. University, Rohtak.

22. Whether 'no-objection' certificate (s) from the co-author (s) have been enclosed ? (attach photocopy of the certificate (s) : Copy of the NOC attached

23. Indicate the complete travel plan from the proposed date and time of departure from the place of working to the conference and back : Travel plan given below

Date	***From***	***To***	***Mode of travel***
13-11-2012	Rohtak – 8 a.m.	New Delhi – 11.00 a.m.	Taxi
13-11-2012	New Delhi – 2 p.m.	Bangkok – 7 p.m.	Air India – AI 332
13-11-2012	Bangkok – 9.30 p.m. TG-122	Chiang Mai – 10.40 p.m.	Thai Airways,
18-11-2012	Chiang Mai – 8.50 p.m.	Bangkok – 10.10 p.m.	Thai Airways, TG-121
19-11-2012	Bangkok – 8.55 a.m.	New Delhi – 11.55 a.m.	Air India, AI-333
19-11-2012	New Delhi – 1 p.m.	Rohtak – 4 p.m.	Taxi

24. Do the conference authorities send the paper for review before accepting it? : Yes

25. Indicate the amount to be paid to the organizers as registration fee (copy of the Registration Form to be enclosed) : 450 USD

    Assistance required from the Commission : FULL

(a) Travel within India to reach the nearest airport : Rohtak to IGI Airport, New Delhi 100 km. (Rs. 1000)

(b) Airfare (both ways) : 29,119 INR (Proof attached)

(c)	Registration fee	:	450 USD *i.e.* $450 \times 56 = 25200$ INR
(d)	Per-diem required (indicate the number of days and the rate)	:	Nov. 13 to Nov. 19 (7 days) *i.e.*, $3000 \times 7 = 21,000$
	Total (in Rs.)	:	**76319 INR**
26.	Has the applicant approached the organizers / any other agency to	:	(does not apply in case of Vice-Chancellors / Commission Members and UGC Officers)
(a)	Waive registration fee?		
(b)	Support air travel?		
(c)	Get the maintenance allowance?		
(d)	Support boarding and lodging?		
(e)	Any other? (specify)	:	NO
27.	If 'YES' to any one of the above items, indicate the latest position and the amount likely to be made available (attach documentary evidence)	:	Not applicable
28.	Has the applicant availed the financial assistance from UGC for attending seminar / conference / symposium etc. in the last 2/3 years prior to the date of the present conference?	:	NO
(a)	If 'YES' give the details in the following table: Name of the Conference Attended Place and dates of	:	Not applicable

	the conference Financial Assistance availed (in Rs.) UGC sanction letter no. with date		
29.	Proposed date of joining the duty in the institution after the conference is over	:	November 20, 2012
30.	Any other information the applicant would like to give in support of the case.	:	If given a chance, it will be very good opportunity for me to showcase the importance of Traditional Indian Medicines to the International Scientific community.

I certify that

(a) The details given above are correct.

(b) If the information supplied is found to be incorrect at a later date, I shall reimburse the entire amount to the Commission.

(c) The amount received will be used for the purpose for which it is requested.

(d) In case financial assistance is received from the organizers or any other agency I shall pay back the amount granted by the Commission.

(e) I shall abide by the decision of the Commission.

Place:

Date:

(Signature of the applicant)
Designation

Appendix - 1

Certificate by Head of the Institution:

I certify that:

(i) The details given by the applicant are correct.

(ii) The applicant has not availed the provision in the last 2/3 years.

(iii) The college has been declared fit to receive financial assistance under section 12(B) of the UGC Act (the letter of inclusion of 12(B) of UGC may be enclosed)

(iv) The applicant has enclosed all the relevant documents.

(v) The information provided in the application is correct.

(vi) In case the institution is found not fit under section 12(B) of the UGC Act even after the approval of the case, the grant will automatically be considered as cancelled.

(vii) In the case of SC/ST/OBC (excluding creamy layer) Teachers, it is certified that they belong to SC/ST/OBC (excluding creamy layer) community.

Signature :

Name :

Designation :

Address :

Office seal :

Date :

SAMPLE PROPOSAL- SEMINAR GRANT

Department of AYUSH
Ministry of Health & Family Welfare
IRCS Building, 1 Red Cross 1 Road,
New Delhi – 110001

Application for grant of financial assistance for organizing Seminars /Symposium /Workshops under Scheme for assistance for Exchange Programme/ Seminars/Workshops on AYUSH.

1. Name of Scientific Association/ Body/ Society/ NGO/ Institution seeking financial assistance :

   Department of Pharmaceutical Sciences, M.D. University, Rohtak, Haryana

2. Full postal address :

   Department of Pharmaceutical Sciences, M.D. University, Rohatk, Haryana-124001, India.

3. Please indicate whether it is the main Society or Chapter/ Unit/Branch of the main Association/Body/Society and status of the organization which is applying :

   Our Organisation, Department of Pharmaceutical Sciences, Maharshi Dayanand University, Rohtak is organising annual convention of Society of Pharmacognosy (Regd.)

4. In case of NGO- Article of Association, bye-laws, audited statement of accounts for last three years, activities and performance report for the last three years, sources and pattern of income and expenditure etc. must be enclosed :

   Not applicable

5. Certificate showing the institution has not obtained or applied for grants for the same purpose or activity from any other

Ministry or Department of the Government of India or the State Government may also be enclosed :

Similar grant has not been applied elsewhere.

6. Topic and subject of Seminar/ Symposium/ Workshops (please be specific) (give brief synopsis) :

TOPIC

PROMOTION AND GLOBALISATION OF INDIAN HERBAL PRODUCTS - PERSPECTIVES AND PROSPECTS

Brief Synopsis:

The use of plants as medicines predates written human history. Ethnobotany (the study of traditional human uses of plants) is recognized as an effective way to discover future medicines. In 2001, researchers identified 122 compounds used in modern medicine which were derived from "ethnomedical" plant sources; 80% of these have had an ethnomedical use identical or related to the current use of the active elements of the plant. Many of the pharmaceuticals currently available to physicians have a long history of use as herbal remedies.

The use of herbs to treat disease is almost universal among non-industrialized societies, and is often more affordable than purchasing expensive modern pharmaceuticals. The World Health Organization (WHO) estimates that 80 percent of the population of some Asian and African countries presently use herbal medicine for some aspect of primary health care. Studies in the United States and Europe have shown that their use is less common in clinical settings, but has become increasingly more in recent years as scientific evidence about the effectiveness of herbal medicine has become more widely available.

The annual turnover of the Indian herbal medicinal industry is about Rs. 2,300 crore with a growth rate of about 10 percent. The export of medicinal plants and herbs from India has increased in the last few years but not significantly as Chinese herbal industry and few other developing countries despite of the fact that India has well-recorded and well-practiced knowledge of traditional herbal medicines. There may be several factors behind this like, lack of well-documented traditional use, single-plant medicines, medicinal plants free from pesticides, heavy metals etc., standardization based on chemical and activity profile and well framed regulatory aspects. But still there is an enormous scope for India also to emerge as a major player in the global herbal product based medicine. Present convention is planned to discuss various issues related to the globalization of Indian Herbal Products and the necessary steps to be taken for the same.

Let us hope that drug manufactured in accordance with principles of Ayurveda, Siddha and Unani will reach new horizons and make them the best in the world if the quality of the herbal drugs is maintained, efficacy would itself be maintained and then there is nothing to stop them from competing with the modern medicine with added advantages of fewer side effects and lower costs.

Objectives

Currently India is having dismal share in Global herbal market because there are no clear cut policies and directions for the cultivation, quality control, manufacturing, R & D and export promotion of Indian herbal products despite of being huge potential. The main objective of this convention is to address and deliberate the various issues related to the promotion and development of Indian herbal products for better acceptance at global level.

Topics to be covered

- The Present Status of Herbal Market in India- Existing Systems, Schemes, Models and Best Practices
- Globalisation of Herbal Products- Scope, challenges, opportunities and development Strategies
- Quality evaluation and value addition of natural products with formulation development, quality control and analysis of natural products.
- Legal policy framework on traditional knowledge protection and IPR issues
- Cultivation of medicinal plants- Policies and viabilities
- Academic and Industrial Herbal Research – Quality analysis and directions for Global acceptance
- Entrepreneurship in Herbal Industry – Scope, Govt initiatives and financial viability
- Regulatory framework- Recent guidelines, lacunas and future needs

7. Date(s) and place of holding Seminar/ Symposium/Workshop :

   21-22 Feburary, 2014 / Department of Pharmaceutical Sciences, M.D. University, Rohtak-124001, Haryana

8.(a) Scientific details of the Seminar/Symposium/Workshop (including various technical sessions). A tentative programme of activities may be supplied:

Day – I (February 21, Friday)

Registration	:	8.00 to 09.30 a.m.
Inauguration	:	09.30 to 11.00 a.m.

Technical session-1

Plenary Lecture-I	:	11.15 to 12.00 p.m.
Plenary Lecture-II	:	12.00 to 12.30 p.m.
Plenary Lecture-III	:	12.30 to 01.00 p.m.

Technical session-2

Plenary Lecture-IV	:	02.00 to 02.30 p.m.
Plenary Lecture-V	:	02.30 to 03.00 p.m.
Short Lectures	:	03.00 to 04.00 p.m.
Poster presentation	:	04.15 to 06.00 p.m.

Day –II (February 22, Saturday)

Technical session-4

Plenary Lecture-VI	:	09.30 to 10.00 a.m.
Plenary Lecture-VII	:	10.00 to 10.30 a.m.
Plenary Lecture-VIII	:	10.30 to 11.00 a.m.
Plenary Lecture-IX	:	11.15 to Tea 12.00 p.m.
Short Lectures	:	12.00 to 1.00 p.m.
Poster presentation	:	02.00 to 03.45 p.m.
Valedictory Function	:	4.00 to 5.00 p.m.

(b) Relevance and important of the topic in the context of priority areas of national health needs:

The standardised herbal products manufactured in quality production houses governed by well formulated regulations is the need of the hour and has total relevance. Also manufacturing of quality herbal products has always a priority of the Govt. Agencies.

(c) Explain briefly as to how the subject of the seminar symposium /Workshop is directly related to dissemination of proven results of AYUSH:

Department of AYUSH has always emphasised on the promotion of herbal drugs and formulations. Also it has promoted the use of technology for improvement of quality of research and herbal products as a whole for better acceptance at both National and International platform. Present programme is mainly focussed on these issues thus directly related to the dissemination of proven results and priorities of AYUSH.

(d) In what way the Seminar/Symposium/Workshop is expected to contribute to the existing knowledge in the field?

This is a unique convention planned on this topic and very less attempts have been made on this core issue. The contents of the event are selected carefully and will be strictly adhered during the convention. It is planned and expected that the key issues as mentioned above will be addressed by leading experts in their respective field. So definitely some important deliberations will come out which will be communicated to the suitable authorities for further development of the herbal sector. Since there will be enough participation of the industry persons and the professional from regulatory affairs apart from leading academicians of the country, this will prove as a suitable platform for future collaborations in the field of manufacturing, export and research avenues.

(e) In case the topic of the Seminar/Symposium/Workshop is the same as in previous years, what is your justification for the same?

No, The topic of the conference is unique and very important.

(f) Has any Chapter/Unit/Branch of the Association/Body/ Society/ NGO received any grant from the Central Government *i.e.*,

Ministry of Health & Family Welfare Department of AYUSH during the last three years for organizing Seminar/ Symposium /Workshop? If so, give details year-wise and quote the relevant reference (letter No. and date), in tabular form under the following heads, along with a report of work done/ summary of achievements and its effect on the society at large :

Not Applicable

(g) If the application is from an institute/ Department, give details regarding collaboration, if any, with particular institute/ Department/ Representative National scientific bodies :

Not Yet.

9.(a) How many delegates are expected to participate? (indicate the number of national and international delegates separately) :

National : 500

International : Nil

(b) How many of the delegates are expected to present papers? (Please give their names, designation and topics). If abstracts have been received, please send copies :

Around 200 delegates are expected to present their papers. About 10 Plenary lectures and 10 short lectures will be delivered by the key persons of the related topics. The announcement is yet to be made so abstracts are yet to be received by us. Details can be provided afterwords.

(c) Please give structure of Seminar, speakers with their topic for each session :

Tentative Speakers for Plenary Lectures

Theme/Topic	*Speaker*
Globalisation of Herbal Products- Scope, challenges, opportunities and development Strategies	xxxxxx
The Present Status of Herbal Market in India-Existing Systems, Schemes, Models and Best Practices	xxxxx
Quality evaluation and value addition of natural products with formulation development, quality control and analysis of natural products.	xxxx
Legal policy framework on traditional knowledge protection and IPR issues	xxxx
Cultivation of medicinal plants- Policies and viabilities	xxxxx
Academic and Industrial Herbal Research – Quality analysis and directions for Global acceptance	xxxx
Entrepreneurship in Herbal Industry – Scope, Govt initiatives and financial viability	xxxx
Regulatory framework- Recent guidelines, lacunas and future needs	xxxxx

(d) To how many delegates is TA/DA offered?

Around 15 delegates will be provided TA/DA.

10.(a) What is the total anticipated expenditure? Please give details under various heads.

a.	TA/DA for Young Scientists (Indian)	50,000/-
b.	TA/DA for Senior Scientists (Indian)	1,00,000/-
c.	Pre-conference printing (Announcements, abstracts, etc.)	50,000/-
d.	Publication of Proceedings	50,000/-
e.	Stationery	30,000/-
f.	Secretarial Assistance	30,000/-
g.	Local Hospitality	3,00,000/-
h.	Misc.	40,000/-
	Grand Total:	**6,50,000/-**

Grant Requested from AYUSH:

At least Rs. 2.5 lakh are requested to be sanctioned from AYUSH to organise this event.

(b) What will be the contribution of the organization?

The organisation will provide proper infrastructure like, Seminar Halls, Furniture, Sound system, Partial transport etc.

11 Details of grants requested / received from other agencies like UGC, INSA,DST,CSIR and ICAR for the proposed Seminar/ Symposium/ Workshop:

Name of Agency	*Grant requested*	*Grant received*	*Items for which*	*Whether U.C.*
		Not any		

12.(a) Income from participants by way of Registration fee etc. :

Approx. 3,00,000/-

(b) Income from other sources : 50,000/-

(By way of Advertisement, sponsorship etc.)

13.(a) Whether report and Utilization Certificate for grant, if any, received earlier, from the Central Government have been submitted?

Not Applicable

(b) Name of the authority who will be responsible for submitting the audited statement of accounts / Utilization Certificate and proceedings/ Report of the Seminar/ Symposium/ Workshop:

Registrar,

Maharshi Dayanand University,

Rohtak.

14.(a) Are the Organizers ready to furnish:

A brief summary report on the Seminar/ Symposium/ Workshop and its impact on participants:

YES

(b) One copy of the proceedings, as and when published:

YES

15. Any other information relevant to the context.

Not any

SAMPLE PATENT FORMATS AND PATENT CERTIFICATE

FORM 1 THE PATENTS ACT 1970 (39 OF 1970) & THE PATENT RULES, 2003 APPLICATION FOR GRANT OF PATENT [See section 7,54 &135 & rule 20(1)	(FOR OFFICE USE ONLY) Application No.: Filing Date: Amount of Fee Paid: CBR No.: Signature:

1. APPLICANT (S)

Name	*Nationality*	*Address*
ZTE Corporation	CN	ZTE Plaza, Keji Road South, Hi-Tech Industrial Park, Nanshan District, Shenzhen City, Guangdong Province 518057, P. R. China

2. INVENTORS (S)

Name	*Nationality*	*Address*
ZHAO, Zhiyong	CN	ZTE Plaza, Keji Road South, Hi-Tech Industrial Park, Nanshan District, Shenzhen City, Guangdong Province 518057, P. R. China
GUO, Feng	CN	ZTE Plaza, Keji Road South, Hi-Tech Industrial Park, Nanshan District, Shenzhen City, Guangdong Province 518057, P. R. China
ZONG, Baiqing	CN	ZTE Plaza, Keji Road South, Hi-Tech Industrial Park, Nanshan District, Shenzhen City, Guangdong Province 518057, P. R. China

3. TITLE OF THE INVENTION:

A method for transmitting multi-wireless-schemes IQ data between a basic band and a radio frequency

4. ADDRESS FOR CORRESPONDENCE OF APPLICANT / AUTHORISED PATENT AGENT IN INDIA

S.P. ARORA Global IPR Law Consultants (India) 6457, Sector C-6 & 7, Vasant Kunj, New Delhi-110070, India	Telephone No: 011-26135156 Fax No.: 011-26897431, Mobile No. 9810604554 E-mail: sparora6457@hotmail.com

5. PRIORITY PARTICULARS OF THE APPLICATION (S) FILED IN CONVENTION COUNTRY

Country	*Application No.*	*Filing Date*	*Name of Applicants*	*Title of Invention*
CN	200710126190.1	June 15, 2007	ZTE Corporation	A method for transmitting multi-wireless-schemes IQ data between a basic band and a radio frequency

6. PARTICULARS OF FILING PATENT COOPERATION TREATY (PCT) NATIONAL PHASE APPLICATION

International application number:	*International filing date as alloted by the receiving office:*
PCT/CN2007/003920	December 29, 2007

7. DECLARATION:

(i) Declaration by the Inventor(s)

I/We, the above named inventor(s) is/are the true & first inventor(s) for this invention and declare that the applicant(s) herein is/are my/our assignee or legal representative.

(1) (a) Date :

(b) Signature(s) of Inventors :

(c) Name(s) ZHAO, Zhiyong

(2) (1) (a) Date :

(b) Signature(s) of Inventors :

(c) Name(s) GUO, Feng

(3) (1) (a) Date :

(b) Signature(s) of Inventors :

(c) Name(s) ZONG, Baiqing

(ii) Declaration by the Applicant(s) in the Convention Country

I/we, the applicant(s) in the convention country declare that the applicant(s) herein is/are my assignee or legal representative.

(a) Date:

(b) Signature(s) :

(c) Name(s) of the Signatory :

(iii) Declaration by the Applicant(s):

- I/we, the applicant(s) are in possession of the above-mentioned invention.

- The complete specification relating to the invention is filed with this application.
- There is no lawful ground of objection to the grant of the Patent to me/us
- The application or each of the applications, particulars of are given in Para 5 was the first application in convention country/countries in respect of my/our invention.
- I/we claim the priority from the above-mentioned application(s) filed in convention country and state that no application for protection in respect of the invention had been made in a convention country before that date by me/us or by any person from which I/ we drive the title.
- My/our application in India is based on international application under patent Cooperation Treaty as mentioned in Para-6.

---

8. FOLLOWING ARE ATTACHMENTS WITH THE APPLICATION:

(a) FORM 2-Complete Specifications, No. of Pages 31 (including Form-2-1 page, Written Description - 20, Claims Pages - 3, Drawings - 6, Abstract - 1) No. of Claims 21 (in duplicate)

(b) Statement and Undertaking on Form 3 (in duplicate)

(c) Declaration as to Inventorship on Form 5 (in duplicate)

(d) WIPO publication – Front Page

(e) Copy of International Search Report

(f) Copy of PCT/IB/304

(g) Copy of International Application Status Report

(h) Certified copy of English Translation of PCT Application

(i) Certified copy of General Power of Attorney

(j) Fee of Rs 13,200/- By Cheque bearing no. 533224 Dated January 5, 2010 On Indian Overseas Bank, Sector-10, Dwarka, New Delhi.

I/We hereby declare that to the best of my /our knowledge, information and belief the fact and matters stated herein are correct and I/We request that a Patent may be granted to me/us for the said invention.

Dated this January 5, 2010

(xxxxxxxx)
Authorized Agent for the applicant
Patent Agent Registration No. In/Pa-389

To
The Controller of Patents
The Patent Office, at New Delhi

Appendix - 3

FORM 2

THE PATENTS ACT 1970
(39 of 1970)
&
THE PATENT RULES, 2003
COMPLETE SPECIFICATION
(See section 10 and rule 13)

1. TITLE OF THE INVENTION:

A Method for transmitting multi-wireless-schemes IQ data between a basic band and a radio frequency

2. APPLICANT (S):

(a) Name : ZTE CORPORATION

(b) Nationality : CN

(c) Address : ZTE Plaza, Keji Road South, Hi-Tech Industrial Park, Nanshan District, Shenzhen City, Guangdong Province 518057, P. R. China

3. PREAMBLE OF THE DESCRIPTION

The following specification particularly describes the invention and the manner in which it is to be performed

FORM 3

THE PATENTS ACT 1970
(39 OF 1970)
&
THE PATENT RULES, 2003
STATEMENT OF UNDERTAKING UNDER SECTION 8
(See section 8, rule 12)

I/We

(a) Name : ZTE CORPORATION

(b) Nationality : CN

(c) Address : ZTE Plaza, Keji Road South, Hi-Tech Industrial Park, Nanshan District, Shenzhen City, Guangdong Province 518057, P. R. China

hereby declare:

(i) that I/We who have made this application No 79/DELNP/2010. Dated January 5, 2010, corresponding to PCT application PCT/CN2007/003920 dated December 29, 2007 alone for the same/substantially same invention, application(s) for Patent in other countries, the particulars of which are given below:

Name of Country	*Application No.*	*Date of Application*	*Status of Application*	*Date of Publication*	*Date of Grant*
China	200710126190.1	June 15, 2007	Published	December 17, 2008	
Europe	07855920.0	Jan. 14, 2010	Published	Mar. 10, 2010	
Europe	12174254.8	June 29, 2012			
Hong Kong	10107271.3	July 29, 2010	Published	Oct. 22, 2010	
PCT	PCT/CN 2007/003920	December 29, 2007	Published	December 18, 2008	WO/2008/151494

(ii) that the rights in the application(s) has/have been assigned to us that we undertake that upto the date of grant of the patent by the Controller, I/We will keep him informed in writing the details regarding corresponding applications for patents filed outside India within six months from the date of filing such application.

Dated this 11th day July, 2012

(xxxxxxx)
Authorized Agent for the Applicant
Patent Agent Registration No. IN/PA-389

To
The Controller of Patents
The Patent Office, at New Delhi

FORM 5

THE PATENTS ACT 1970
(39 OF 1970)
&
THE PATENT RULES, 2003
DECLARATION AS TO INVENTORSHIP
[See section 10(6) and Rule 13(6)]

I/We

1. NAME OF APPLICANTS

(a)	Name	:	**ZTE CORPORATION**
(b)	Nationality	:	CN
(c)	Address	:	ZTE Plaza, Keji Road South, Hi-Tech Industrial Park, Nanshan District, Shenzhen City, Guangdong Province 518057, P.R. China

hereby declare that the true and first inventor(s) of the invention disclosed in the complete specification filed in pursuance of my /our application numbered dated January 5, 2010 corresponding to PCT Application No. **PCT/CN2007/003920**, is/are

2. INVENTORS (S)

Name	*Nationality*	*Address*
ZHAO, Zhiyong	CN	ZTE Plaza, Keji Road South, Hi-Tech Industrial Park, Nanshan District, Shenzhen City, Guangdong Province 518057, P.R. China
GUO, Feng	CN	ZTE Plaza, Keji Road South, Hi-Tech Industrial Park, Nanshan District, Shenzhen City, Guangdong Province 518057, P.R. China
ZONG, Baiqing	CN	ZTE Plaza, Keji Road South, Hi-Tech Industrial Park, Nanshan District, Shenzhen City, Guangdong Province 518057, P.R. China

3. DECLARATION TO BE GIVEN WHEN THE APPLICATION IN INDIA IS FILED BY THE APPLICANT (S) IN THE CONVENTION COUNTRY

We ,the applicant(s) in the convention country hereby declare that our right to apply for a patent in India is by way of assignment from the true and first inventor(s)

Dated this January 5,2010

(xxxxxxx)

Agent for the Applicant

Patent Agent Regn No. xxx

To

The Controller of Patents

The Patent Office, at New Delhi

Appendix - 3

FORM 1 **THE PATENTS ACT 1970** **(39 OF 1970)** **&** **THE PATENT RULES, 2003** **REQUEST/EXPRESS REQUEST FOR EXAMINATION OF APPLICATION FOR PATENT** **[See Section 11B and Rules 20(4)(ii),24B(1)(i)]**	(FOR OFFICE USE ONLY) RQ No.: Filing Date: Amount of Fee Paid: CBR No. Signature

1. APPLICANTS

(a) Name : ZTE CORPORATION

(b) Nationality : CN

(c) Address : ZTE Plaza, Keji Road South, Hi-Tech Industrial Park, Nanshan District, Shenzhen City, Guangdong Province 518057, P.R. China

2. STATEMENT IN CASE OF REQUEST FOR EXAMINATION MADE BY THE APPLICANT (S)

I/We hereby request that my/our application for patent no. 79/DELNP/2010 dated January 5, 2010 for the invention titled "A METHOD FOR TRANSMITTING MULTI-WIRELESS-SCHEMES IQ DATA BETWEEN A BASIC BAND AND A RADIO FREQUENCY" be examined under Section 12 and 13 of the Act

3. ADDRESS FOR SERVICE:

GLOBAL IPR LAW CONSULTANTS (INDIA)
6457, C-6 & 7, Vasant Kunj, New Delhi-110070
Tele No.: 011-26135156, 9810604554; Fax No.: 011-26897431
E-mail: sparora6457@hotmail.com

Dated this 7th day of January , 2010

Signature

(xxxx)

Agent for the Applicant

Patent Agent No.IN/PA- xxx

To

The Controller of Patents

The Patent Office at New Delhi

सत्यमेव जयते

INTELLECTUAL PROPERTY INDIA

GOVERNMENT OF INDIA
PATENT OFFICE
INTELLECTUAL PROPERTY BUILDING
Plot No. 32, Sector-14,Dwarka
New Delhi - 110 078

Tel No. (091)(011) 28034304-06,22
Fax No. 011 28034301,28034302
E-mail : delhi-patent@nic.in
Web Site : www.ipindia.nic.in

Letter No.:-ELEC/2013/ 1661

Date : 28/01/2013

To,
GLOBAL LAW CONSULTANTS (INDIA)
6457, SECTOR C 6 & 7, VASANT KUNJ,
NEW DELHI-110070, INDIA.

SUB : First Examination Report

APPLICATION NUMBER	:	6140/DELNP/2005
DATE OF FILING	:	29/12/2005
DATE OF REQUEST FOR EXAMINATION	:	12/07/2006
DATE OF PUBLICATION	:	11/07/2008

a) With reference to the RQ No. 5902/RQ-DEL/2006 Dated 12/07/2006 in the above mentioned application for Grant of Patent , Examination has been conducted under Section 12 and 13 of the Patents Act 1970 , The following objections are hereby communicated.

b) Objections :

1 Claim(s) 1-8 fall(s) within the scope of such clause (k) of section 3.

2 Claims do not sufficiently define the invention. Characterize all essential novel features of the invention in main claim, where lies novelty in claims, pin point clearly, reference numerals should be given in claims to enhance intelligibility of the claims.

3 Details regarding the search and/or examination report including claims of the application allowed, as referred to in Rule 12(3) of the Patent Rule, 2003, in respect of same or substantially the same invention filed in all countries outside India along with appropriate translation where applicable, should be submitted within a period of Six months from the date of receipt of this communication as provided under section 8(2) of the Indian Patents Act.

4 Details regarding application for Patents which may be filed outside India from time to time for the same or substantially the same invention should be furnished within Six months from the date of filing of the said application under clause (b) of sub section(1) of section 8 and rule 12(1) of Indian Patent Act.

5 Necessary figure may be indicated in the abstract and abstract should be prepared in accordance with the instructions contained in the Rule 13{7(d)} of the Patent Rules, 2003(as amended in 2006). Reference numerals wherever necessary in abstract may be given.

6 The Drawings referred to in the specification should be prepared inaccordance with the instructions contained in the Rule 15 of the Patent Rules, 2003(as amended in 2006).

c) You are requested to comply with the objections by filing your reply by way of explanation and/or amendments within 12 months from the date of issue of FER failing which your application will be treated as "Deemed to have been abandoned" under section 21(1) of the Act. The last Date is 28/01/2014.

d) You are advised to file your reply at the earliest so that the office can further proceed with application and complete the process within the prescribed period.

28/01/2013

(S N Sav)
Asst. Controller of Patents & Designs

NOTE : All Communications to be sent to the Controller of Patents at INTELLECTUAL PROPERTY BUILDING Plot No. 32, Sector-14,Dwarka New Delhi - 110 078.

Appendix - 3

GOVERNMENT OF INDIA
PATENT OFFICE
INTELLECTUAL PROPERTY BUILDING
Plot No. 32, Sector-14, Dwarka, New Delhi - 110 075

Tel No. (091)(011) 28081922 - 25
Fax No. 011 28081940,20
E-mail : delhi-patent@nic.in
Web Site : www.ipindia.nic.in

1102

No. 261387

Dated the:24/06/2014

To

S.P. ARORAGLOBAL IPR
LAW CONSULTANTS
(INDIA)6457, SECTOR C 6
& 7, VASANT KUNJ, NEW
DELHI-110070, INDIA.

BY REGISTERED A.D

REF :- Patent No.261387 (5674/DELNP/2007), Dated 23/07/2007(Ante-Dated :19/01/2006) Granted On Dated 24/06/2014

This is to state that a patent has been granted on the above-mentioned application and that the grant of the patent has been recorded in the Register of Patents on the 24/06/2014. The said patent is enclosed herewith.

The payment of renewal fee is required to be made at this office within three(3) months from the aforesaid date of recording according to the proviso in sub-section(4) of Section 142 of the Patents Act,1970, as Amended by the Patent(Amendment)Act,2005/ Patent Rule, 2003 as Amended by Patent(Amendment) Rules,2006.

The renewal fee schedule has been given at the back of the Letter of Patent

Encl: As above

(B.P.Singh)
Deputy Controller of Patents & Designs

क्रमांक : | 011 | 022258
Sl. No. :

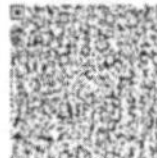

भारत सरकार
GOVERNMENT OF INDIA
पेटेंट कार्यालय
THE PATENT OFFICE
पेटेंट प्रमाणपत्र
Patent Certificate
(Rule 74 of Patents Rules)

Patent Number	:	261387
Application Number	:	5674/DELNP/2007
Date of Filing	:	19/01/2006
Patentee	:	BEHR PROCESS CORPORATION

It is hereby certified that a patent has been granted to the patentee for an invention entitled APPARATUS AND METHOD FOR COLOR SELECTION AND COORDINATION SYSTEM as disclosed in the above mentioned application for the term of 20 years from the 19 day of JANUARY 2006, in accordance with the provisions of the Patent Act 1970.

Controller of Patents

Controller General of Patents,
Designs & Trademarks

Date of Grant:24/06/2014

Note: The fees for renewal of this patent, if it is to be maintained, will fall/has fallen due on 19 day of JANUARY 2008 and on the same day in every year thereafter.

Appendix - 3

READER'S ASSESSMENT

Communication Skills in Scientific Research

By

Dr. Munish Garg

Dr. Chanchal Garg

Please put tick mark in appropriate box and mail to the author or send your views for improvement at e mail: mgarg2006@gmail.com

1.	Contents	Excellent	Good	Satisfactory
2.	Presentation	Superb	Nice	Upto the mark
3.	Get up	Excellent	Good	Satisfactory
4.	Language	Lucid	Effective	Inconsistent
5.	Overall Rating	Excellent	Above average	Average

Suggestions for improvement:

....

....

....

....

....

....

Subject Index

G

H

I

K

L

M

N

O

P

R